THE IMPACT OF NEAR-INFRARED SKY SURVEYS
ON GALACTIC AND EXTRAGALACTIC ASTRONOMY

ASTROPHYSICS AND SPACE SCIENCE LIBRARY

VOLUME 230

THE IMPACT OF NEAR-INFRARED SKY SURVEYS ON GALACTIC AND EXTRAGALACTIC ASTRONOMY

Proceedings of the 3rd EUROCONFERENCE
on Near-Infrared Surveys held at Meudon Observatory, France,
June 19–20, 1997

Edited by

N. EPCHTEIN

Observatoire de la Côte d'Azur, Nice, France,
and
Département de Recherche Spatiale,
Observatoire de Paris, France

SPRINGER-SCIENCE+BUSINESS MEDIA, B.V.

Library of Congress Cataloging-in-Publication Data

ISBN 978-94-010-6110-0   ISBN 978-94-011-5026-2 (eBook)
DOI 10.1007/978-94-011-5026-2

Printed on acid-free paper

TABLE OF CONTENTS

VI-Miscellaneous

FOREWORD

Thirty years after the pioneering enterprise of Neugebauer and collaborators, the astronomical sky is surveyed anew in the near infrared range with a gain in sensitivity greater than 4 orders of magnitude. Data have started to come through the "pipelines" routinely and at the turn of the century, the astronomical community will be provided with immense databases that will eventually contain accurate positions and 3- colour (or even 4-colour after merging DENIS and 2MASS data) photometry for hundreds million of stars and millions of galaxies.

This fantastic harvest of data will eventually result of the huge effort that has been deployed on both sides of the Antlantic to promote 2 major projects, 2MASS and DENIS.

The basic aim of this 3rd Euroconference (and last in the series), was to put in close contact the scientific teams of 2MASS and DENIS in order to present and discuss the first significant results of the two surveys and to start promoting possible future plans of collaboration. It was preceded by a workshop, with a smaller attendance, sponsored by a NSF-CNRS action, that was dedicated to the technical aspects of the data processing and database implementation and management. The Euroconference itself, although short (2 days) has been rather dense and animated, and real advances in various fields were presented that promise a final great success.

After several months of operations in *proto-phase* or in routine phase, and having archived a good deal of data, we are now in a better position to start answering some of the key questions that were addressed at the beginning of the projects. What are we going to learn with the 2 micron survey data ? What will be the size of the final catalogues (stars and galaxies)? How homogeneous, complete, and reliable will they be ? What are the most appropriate algorithms for image processing, source extraction, star/galaxy separation, photometric calibration of point sources and low surface brightness objects. Finally, how to manage to the best the astrophysical exploitation and follow–up of the surveys, while 8–10 meter class telescopes are flourishing around the world, while ISO is about to end up after a very successful mission, and while future space missions such as SIRTF are in preparation phase. Some answers are extremely exciting. We will eventually find hundreds of brown dwarfs and thousands of very low mass stars, improve the local luminosity function and several basic parameters of the galactic structure, probe the stellar populations of the Magellanic clouds and of the bulge. The largest photometric catalog of galaxies will be produced, superseding in reliability and photometric accuracy the catalogs based on digitised photographic material. This will lead to unprecedented statistical works on the local structure of the Universe and the census of galaxies in the zone of avoidance, just to cite a few exemples.

This conference has, indeed, been very fruitful in exchanging information between the American and European teams. It has been also a good opportunity to compare our working technics and methods, and to note that the European Science needs to be better organised to manage middle-size projects that are not fully sponsored by official institutions such as ESO.

The European Commission (DGXII) that supported this cycle of EU-ROCONFERENCE and the early phase of DENIS is thanked for its invaluable contribution. I sincerely regret that it did not pursue its support while the programme is entering its active phase of scientific exploitation.

It is a great pleasure for me to warmly thank all the participants to this conference, the Scientific and Local Organising Committees, and especially those who have presented excellent talks and provided me with outstanding written contributions. I am also greatly indebted to Eric Copet, Christine Denis, Constance Imad, Jacqueline Thouvay, Jean–Jacques Poisot and Suzanne Berton, from Observatoire de Paris and Josette Schmidt from Institut d'Astrophysique de Paris for the fine organisation of the logistics of the conference sessions and dinner.

Finally, I thank Alain Omont for having taken the responsibility of this cycle of EUROCONFERENCE during the last 3 years.

Nicolas Epchtein

LIST OF PARTICIPANTS

Alard, Christophe
alard@iap.fr
Observatoire de Paris, DASGAL   France

Alvarez, Pedro
alvarez@graal.univ-montp2.fr
GRAAL Montpellier   France

Bartlett, James
bartlett@cdsxb6.u-strasbg.fr
Observatoire de Strasbourg   France

Barucci, Maria-Antonella
barucci@obspm.fr
Observatoire de Paris, DESPA   France

de Batz, Bertrand
debatz@obspm.fr
Observatoire de Paris, DASGAL   France

Baudrand, Anne
baudrand@obspm.fr
Bureau des Longitudes, Paris   France

Becker, Robert
bob@igpp.llnl.gov
University of California-Davis   USA

Berend, Zsolt
berend@ogyalla.konkoly.hu
Konkoly Observatory, Budapest   Hungary

Bertin, Emmanuel
ebertin@eso.org
ESO, Garching   Germany

Bienaymé, Olivier
bienayme@ cdsxb6.u-strasbg.fr
Observatoire de Strasbourg   France

Cambrésy, Laurent
cambresy@obspm.fr
Observatoire de Paris, DESPA   France

Carpenter, John M.
carp@pegasus.ifa.hawai.edu
University of Hawaï   USA

Copet, Eric
copet@obspm.fr
Observatoire de Paris, DESPA   France

Cioni, Maria-Rosa
cioni@strw.leidenuniv.nl
Sterrewacht Leiden   The Netherlands

Cornelisse, Remon
cornelisse@ strw.leidenuniv.nl
Sterrewacht Leiden   The Netherlands

Cutri, Roc
roc@ipac.caltech.edu
IPAC/CALTECH   USA

Delfosse, Xavier
delfosse@gag.observ-gr.fr
Observatoire de Grenoble   France

Deul, Erik
deul@strw.leidenuniv.nl
Sterrewacht Leiden   The Netherlands

Epchtein, Nicolas
epchtein@obspm.fr
Observatoire de Paris, DESPA   France

Forveille, Thierry
forveille@gag.observ-gr.fr
Observatoire de Grenoble   France

Garzón, Francisco
fgl@ll.iac.es
Instituto de Astrofisica de Canarias   Spain

Gonzalez-Solares, Eduardo
eglez@ll.iac.es
Instituto de Astrofisica de Canarias   Spain

Groenewegen, Martin A.T.
groen@MPA-Garching.MPG.de
MPI für Astrophysik, Garching   Germany

Hammersley, Peter *plh@iac.es*	Instituto de Astrofisica de Canarias	Spain
Holl, Andras *holl@ogyalla.konkoly.hu*	Konkoly Observatory, Budapest	Hungary
Hron, Josef *hron@astro.ast.univie.ac.at*	Institüt für Astronomie Wien	Austria
Jarrett, Tom H. *jarrett@ipac.caltech.edu*	IPAC/CALTECH	USA
Kimeswenger, Stefan *Stefan.Kimeswenger@uibk.ac.at*	Universität Innsbruck	Austria
Kraan-Korteweg, Renée *kraan@norma.astro.ugto.mx*	Observatoire de Paris, DAEC	France
Lafon, Jean-Pierre *lafon@obspm.fr*	Observatoire de Paris, DASGAL	France
Lopez-Corredoira, Martin *martinlc@iac.es*	Instituto de Astrofisica de Canarias	Spain
Loup, Cécile *cloup@eso.org*	Institut d'Astrophysique de Paris	France
Mamon, Gary A. *gam@iap.fr*	Institut d'Astrophysique de Paris	France
Mobasher, Bahram *b.mobasher@ic.ac.uk*	Imperial College, London	UK
Montmerle, Thierry *montmerle@.cea.fr*	SAp/CEA, Saclay	France
Omont, Alain *omont@iap.fr*	Institut d'Astrophysique de Paris	France
Paturel, Georges *patu@adel.univ-lyon1.fr*	Observatoire de Lyon	France
Price, Stephan *price@plh.af.mil*	Air Force Phillips Lab., Hanscom, Mass.	USA
Rosenberg, Jessica *rosenber@fcrao2.phast.umass.edu*	University of Massachusetts, Amherst	USA
Ruphy, Stéphanie *stephanie.ruphy@larecherche.fr*	Observatoire de Paris, DESPA	France
Schneider, Stephen E. *schneide@wilt.phast.umass.edu*	University of Massachusetts, Amherst	USA
Schultheis, Mathias *schultheis@astro.ast.univie.ac.at*	Institüt für Astronomie Wien,	Austria
Simon, Guy *Guy.Simon@obspm.fr*	Observatoire de Paris, DASGAL	France
Skrutskie, Michael *skrutski@north.phast.umass.edu*	University of Massachusetts, Amherst	USA
Unavane, Mukund *munavane@ast.cam.ac.uk*	IoA, Cambridge	UK
Van Driel, Willem *vandriel@mesioq.obspm.fr*	Observatoire de Paris, Nançay	France
Van Eck, Sophie *svaneck@astro.ulb.ac.be*	Inst. Astron. Astrophys., Bruxelles	Belgique
Vauglin, Isabelle *isa@altair.univ-lyon1.fr*	Observatoire de Lyon	France

I- Infrared Survey Status

THE DEEP NEAR INFRARED SURVEY OF THE SOUTHERN SKY (DENIS):

Progress report, and scientific results overview

N.EPCHTEIN

Observatoire de la Côte d'Azur
Département Fresnel, BP 4229, F06304 Nice cedex
and DESPA, Observatoire de Paris

Abstract. I present the status of achievement of DENIS operations and data processing in November 1997, and summarise the main scientific results obtained so far based on the analysis of the first set of observations. Prospects for the coming years are briefly outlined.

1. Introduction

Although 2MASS and DENIS have followed different development plans, they are basically aimed at similar objectives: mapping the all (southern, in the case of DENIS) –sky primarily at 2 micron (filter K_s), and in 2 accompanying wavelengths, Gunn-I and J for DENIS, J and H for 2MASS. This difference in spectral coverage makes each project slightly better adapted to investigate different astrophysical problems. DENIS is better suited to pick up cool isolated red (and brown) dwarfs, while 2MASS is better adapted to probing star forming regions, since highly extinguished objects are generally not detectable in the I band. In addition, the slightly higher sensitivity of 2MASS in K_s, obtained thanks to a larger telescope aperture and a fully cooled instrument, will undoubtedly provide an advantage in the detection of galaxies and low surface brightness objects.

2. Performances of DENIS

The main specifications of DENIS have been described several times (e.g. Epchtein, 1997a; Epchtein, 1997b; Copet 1996; Ruphy 1996) and are just briefly outlined here. The DENIS survey aims at covering the all–southern sky ($-88°$ to $+2°$ of declination) simultaneously in the I, J, K_s bands at

3

N. Epchtein (ed.),
The Impact of Near-Infrared Sky Surveys on Galactic and Extragalactic Astronomy, 3-9.
© *1998 Kluwer Academic Publishers.*

arcsecond resolution using the ESO 1 meter telescope at La Silla, Chile. Mapping of the sky is performed by scanning the telescope along strips of 30° in declination and 12′ wide in RA in a step–and–stare mode. Each elementary image of 12′ × 12′ is acquired in ≈ 10s. The main characteristics and performances are summarised in Table 1.

TABLE 1. DENIS characteristics and performances

Channel	I	J	K_s
Central wavelength			
(μm)	0.8	1.25	2.15
Arrays	CCD Tektronix	NICMOS3	NICMOS3
Size of the arrays	1024 × 1024	256 × 256	256 × 256
Pixel size (*arcsec*)	1	3	3
Exposure time (*second*)	9	10	10
Limiting mag.			
(point source, 3σ)	18.5	16.5	14.0
Saturation (mag.)	9.5	8.5	6.5

Several improvements of the focal instrument have been designed and set up in 1997. The CCD dewar has been replaced by a new one with a much longer hold-time (more than 32 hours), and most importantly, an air-conditioning system of the focal instrument has been installed. The aim of this device is to reduce the thermal emission background of the *warm* optics and thus to improve the sensitivity of the instrument by at least 0.5 magnitude in the K_s band in Summer time. The temperature of the instrument is stabilized around 5°C by blowing a continuous flow of dry and cool air in an envelope encompassing the whole focal box (Fig 1). The thermal background emission is now independent of the variations of the room temperature. Background emission is reduced by a factor of ≈ 2 with respect to the former worse conditions. The homogeneity of the survey has now considerably improved.

3. Present status of the observations

Since the beginning of the operations, more than 2000 strips have been scanned and reduced, covering about 30 % of the objective. Fig 2 shows the progression of the survey during the first 18 months of operations. The large plateau in southern Winter 1997 corresponds to the interruption of the survey caused by the mechanical modifications described above and a period of particularly bad meteorological conditions. Not all the data are

Figure 1.   The DENIS instrument in its new thermalization envelope

good enough to pass the quality criteria and it is expected that a fraction of the strips will have to be reobserved.

4. Status of the Data Processing

The data processing is performed routinely in the 2 data analysis centres (DACs) at Paris (PDAC) and Leiden (LDAC), according to the data stream displayed in Fig. 4 of Epchtein (1997a). The description of the pipelines has been detailed in the previous *Euroconferences* by Borsenberger (1997) and Deul *et al.* (1995) and are not repeated here. Most of the images taken during the *protosurvey* period and the first year and half of operations have been processed, archived, and delivered to LDAC for source extraction. Information concerning the DENIS database status at PDAC can be retrieved

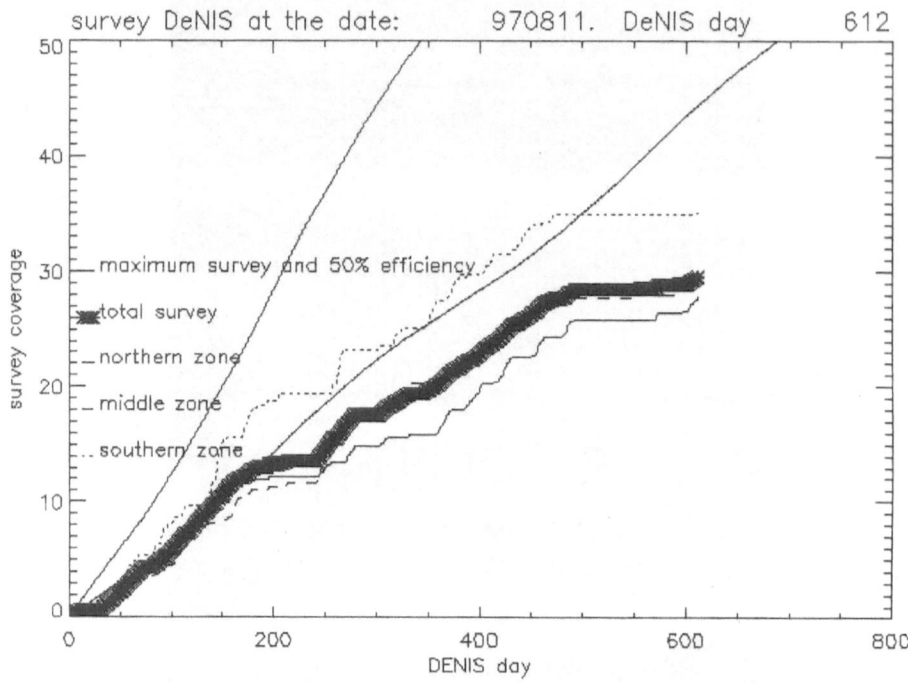

Figure 2. Status of the observations in August 1997(from P. Fouqué)

on the website:

`http://www-denis.iap.fr`

At this time are available, the database containing the history of the observations, (called *FOURBI*) and several utilitary routines that allow, for instance, to check if a given area of the sky has been surveyed and where the relevant data are stored. Access to the database is provisionally restricted to DENIS consortium members, but will be opened soon for general release. Image data are stored in a jukebox of 52 DLTs which contain more than 1000 Gbytes of data (some 25 % of the southern sky).

LDAC has started to perform routinely source extraction and calibration only since September 1997, and a still limited number of strips (a hundred or so as of Nov. 1997) have been processed. Lists of sources per strip can be sorted out (with the same access restriction as for PDAC) at the following web address:

`http://www.strw.leidenuniv.nl/denis/index.html`

5. Science with DENIS, an overview

This Euroconference is essentially dedicated to presenting the scientific analysis of the first data sets of the 2–micron surveys, and the purpose of this book is to summarize these first results, and to try to prepare the best exploitation and follow–up observations of the huge amount of new data soon to come. The main areas in which significant results have been obtained so far are the regions of star formation, the galactic structure, the low mass stars and brown dwarfs and the galaxies. In this section, I briefly mention some of the most prohement results obtained with DENIS and refer to papers included in the present book for further details.

5.1. STAR FORMING REGIONS

The near-infrared range is one of the most appropriate to detect young stellar objects inside or close to their parental molecular cloud and to derive the luminosity function, as well as the variations of the dust extinction (see Carpenter in this volume). For the first time, square degrees of regions of intense stellar formation activity such as the Orion, Chameleon and ρ Ophiuchus molecular clouds are mapped in this spectral range providing panoramic overviews and very large samples of YSO candidates that will lead to improve the initial luminosity and mass functions, as well as the possible variations of the star forming rate within the cloud.

A DENIS/ISO collaboration has begun which is aimed at exploring the most nearby regions of star formation. Persi *et al.* (this volume) present the first cross–identified ISOCAM/DENIS data in the Chameleon I cloud. Cambrésy (this volume) analysing the J counts across the Chameleon cloud made a detailed mapping of the extinction in this region and has discovered a number of new faint YSOs, that are likely to be classical T Tau candidates (Cambrésy et al., 1998).

5.2. STAR COUNTS AND GALACTIC STRUCTURE

Massive star counts in the NIR allow exploring the distribution of the stellar populations in highly extinguished regions, and probing the basic parameters that describe the structure of the Milky Way (e.g., Bienaymé, this volume; Robin, 1997). The immense gain with respect to optical surveys is a consequence of the good transparency of the interstellar dust at 2 μm, that allows to unveil hidden populations of the Galactic disc and bulge.

The combination of the DENIS near–IR and mid–IR observation provided by ISO (7 and 15 μm) in the framework of the ISOGAL survey (Omont *et al.*, this volume) allows a good discrimination between the various species of objects, and in particular easily breaks out AGB stars and YSOs.

Ruphy (1996), using the first release of DENIS data has studied the galactic anticentre direction and analysed strips in these directions. Althouh, in most directions, infrared star counts and models (Robin, 1997; Cohen, 1997 and ref. therein) of stellar populations are in good agreement, toward the anticentre there are significant discrepancies with all models. There are fewer stars observed than the models predict (Ruphy et al.,1996). Moreover, colour diagrams have been used to attempt to separate giant and dwarf populations. An excellent separation is obtained at low latitudes. Based on a comparison between DENIS star counts and models, new revised values of the scale length of the late type giant stars and of the cut–off distance of the Galaxy have been derived and an investigation on the stellar content and shape of the ring at 4 kpc has been initiated (Ruphy et al., 1997).

A full coverage of the Magellanic Clouds is essentially completed that will lead to the construction of an exhaustive catalogue of mass–losing AGB stars and allow an easy discrimination between their various species (Carbon/Oxygen rich star separation) (see Loup et al., this conference)

5.3. RED AND BROWN DWARFS

One of the most spectacular results of DENIS in 1997 has been the discovery of the first isolated brown dwarf (see Forveille et al., this volume). This exciting result, based on the analysis of about 1 % of the DENIS data, is extremely promising for the future, since hundreds of brown dwarfs and thousands of M dwarfs later that M7 will come out of DENIS (and 2MASS). After the completion of the surveys, the first statistically significant catalog of nearby late red dwarfs and brown dwarfs will be available and will usefully complete the former catalogs towards the low end of the luminosity fcunction. The cross–identification of the DENIS and 2MASS observations made at different epochs will be useful to measure large proper motions.

6. Extragalactic

Analyses of the first DENIS and 2MASS data show that the 2 surveys will have a considerable impact on statistical extragalactic atstronomy thanks to the production of huge complete and reliable catalogs providing linear photometry. This last point is obviously a great advantage compared to catalogs based on digitized photographic plate catalogs such as APM or COSMSOS. Clearly, 2MASS thanks to a better sensitivity in K_s, a better homogeneity of the PSF across the frame, and a larger sky surface coverage will produce a much larger K_s catalog of galaxies than DENIS will do (more than 1 million in the all-sky compared to a few tens of thousands in the Southern Sky). One advantage of DENIS, however, is its very sensitive I channel with a 1 arcsec resolution which makes easier the star/galaxy separation. A million galaxies are expected in that band.

7. The future

There is no doubt that the case for a 2-micron all sky survey that were made out almost ten years ago to support the project were good. Most of the objectives will be reached. In the coming years, we will be provided with the largest photometric and astrometric star catalog (one billion star with position better than 1 arcsec and photometry better than 10 %) ever produced, and a large, homogeneous and reliable catalog of galaxies . An unbiased sample of hundreds of brown dwarfs will be probably available for the first time, several large molecular clouds will reveal all their content of relatively young embedded objects (surveys at longer wavelengths, partly done by ISO will be necessary to pick up even younger objects). It is worth mentionning that this extraordinary harvest of data will have been obtained with really little money and small (old–fashion for DENIS) telescopes. I do hope that for the maximum benefit of both surveys, their databases will be easily interconnected, and look forward to prepare the next all–sky survey in the still uncovered spectral domain that spans between 2 and 10 microns.

References

Cambrésy L., Epchtein N., Copet E., de Batz B., Kimeswenger S., Le Bertre T., Rouan D., Tiphène D. (1996) *A&A*, **Vol. no. 265**, pp. 145–332

Cambrésy L., Copet E., Epchtein N. (1998) *A&A*, in press

Borsenberger J., (1997), Proc. of the 2nd Euroconference on *The impact of Large Scale near-infrared sky surveys*, Puerto de la Cruz, Spain, eds. F. Garzon, N. Epchtein, A. Omont, W.B. Burton, P. Persi, Kluwer ASSL series **vol. no. 210**, pp 181–186

Deul E. (+17 authors) (1995) Proc. 1st Euroconference on *Near-Infrared Sky Surveys*, San Miniato, Italy in *Mem. Soc. Astron. Ital.* **vol. no. 66**, pp 549–566

Copet E., (1996) *Thèse de Doctorat, Université de Paris 6,*

Epchtein N. (+ 47 authors)(1997a) *The Messenger*, **Vol. no. 87**, pp. 27–34

Epchtein N. (1997b) Proc. of the 2nd Euroconference on *The impact of Large Scale near-infrared sky surveys*, Puerto de la Cruz, Spain, eds. F. Garzon, N. Epchtein, A. Omont, W.B. Burton, P. Persi, Kluwer ASSL series **vol. no. 210**, pp 15–24

Robin A.(1997) Proc. of the 2nd Euroconference on *The impact of Large Scale near-infrared sky surveys*, Puerto de la Cruz, Spain, eds. F. Garzon, N. Epchtein, A. Omont, W.B. Burton, P. Persi, Kluwer ASSL series **vol. no. 210**, pp 57–61

Ruphy S. (1996) *Thèse de Doctorat, Université de Paris 6,*

Ruphy S., Robin A.C., Epchtein N., Copet E., Bertin E.,Fouqué P. , Guglielmo F. (1996) *A&A*, **Vol. no. 313**, pp. L21–L24

Ruphy S., Epchtein N., Cohen M., Copet E., de Batz B., Borsenberger J., Fouqué P., Kimeswenger S., Lacombe F., Le Bertre T., Rouan D., Tiphène D., (1997) *A&A*, **Vol. no. 326**, pp. 597–607

THE TWO MICRON ALL SKY SURVEY (2MASS): STATUS REPORT NOVEMBER 1997

M.F. SKRUTSKIE

University of Massachusetts
Department of Physics and Astronomy
Box 3-4525, Amherst, MA 01003

Abstract. The Two Micron All Sky Survey (2MASS) began regular survey observations at the Northern Hemisphere 2MASS facility at Mt. Hopkins, Arizona in June 1997. At the time of this writing approximately 10% of the entire sky has been observed. The Southern Hemisphere facility at Cerro Tololo is scheduled to begin taking data in February 1998. 2MASS hopes to release its first increment of data to the community in the first half of 1999.

1. Introduction

The Two Micron All Sky survey will map the entire celestial sphere in the near-infrared J ($1.13 - 1.37\mu$m), H ($1.50 - 1.80\mu$m), and K_s ($2.00 - 2.32\mu$m) bands through the year 2000. Table 1 summarizes the general survey characteristics. One survey facility, which includes a dedicated 1.3m telescope and a three-color infrared camera, began full operations at Mt. Hopkins, Arizona in June 1997. A second facility will come on line in the Southern Hemisphere at Cerro Tololo, Chile in early 1998. On a typical night one of these telescopes automatically observes approximately 60 square degrees of survey area, calibrating once an hour throughout the night. The Infrared Processing and Analysis Center (IPAC) has developed the 2MASS Production Processing System which, running on a Sun Enterprise server, can process an entire night's data to photometrically and astrometrically calibrated point and extended source lists and finished Atlas Images in less than 24 hours. The prototype version of this pipeline currently monitors observatory and camera health and has demonstrated that the survey requirements outlined in Table 1 can be met. IPAC is currently developing the final production version of the pipeline which will begin operating in

N. Epchtein (ed.),
The Impact of Near-Infrared Sky Surveys on Galactic and Extragalactic Astronomy, 11-15.
© *1998 Kluwer Academic Publishers.*

early 1998. The production pipeline will generate the initial 2MASS data catalogs and images which are targeted for general release in the first half of 1999. 2MASS plans to release data incrementally throughout the life of the survey, typically within 12-18 months of the date of observation. At the completion of survey operations in the year 2000 all data will be fully reprocessed to produce a final uniform set of catalogs and images. The final 2MASS point source catalog, which will have a data volume of ~300GBy, will contain approximately 300 million entries. The 2MASS extended source catalog will consist of an extended source list containing about 1 million entries (100 MBy) and postage stamp images of all extended sources (10 Gby). Finally, 2MASS will produce an image atlas (1″ pixels) of the entire sky. The uncompressed full resolution image data will span 14 Tby. A more manageable ~20-times lossy compressed version of the images will be available for general distribution.

TABLE 1. 2MASS Characteristics

Property	Requirement
Arrays	256×256 NICMOS3 (HgCdTe)
Pixel Size	2.0″
Wavebands	J, H, and K_s (2.00 − 2.32μm)
Telescopes	1.3-meter Equatorial Cassegrain
Integration time	6 × 1.3s/frame = 7.8s total
Sensitivity (3σ)	17.1, 16.4, 15.6 mag for J, H, and K_s
Photometric Accuracy	5% for bright sources
Photometric Uniformity	4% over the sky
Positional Accuracy	0.5″
Completeness / Reliability at 10σ	0.99 / 0.9995

The 2MASS observing hardware has been described in other contributions (Skrutskie et al. 1997, Milligan et al. 1996). Current details about the project's hardware/software development and status are available at the project's websites -
`http://pegasus.phast.umass.edu/` and
`http://www.ipac.caltech.edu/2mass/`
This contribution focuses on the project's sky coverage and planned data releases during the survey lifetime. Since the sky coverage proceeds in a systematic fashion one can approximate the portion of sky which will be released as a function of time.

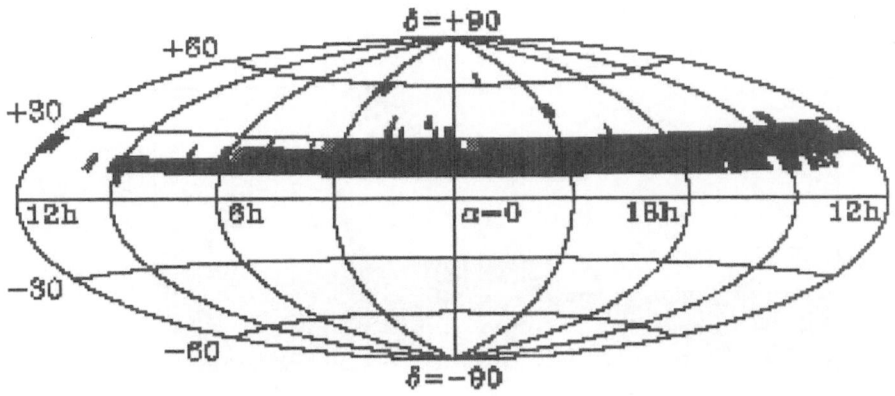

Figure 1.   2MASS Northern Hemisphere sky coverage through November 21, 1997 in equatorial coordinates.

2. Observing Strategy/Schedule

Figures 1 and 2 show the 2MASS sky coverage as of November 21, 1997. Only the Northern Hemisphere facility was in operation during this period. Northern Hemisphere sky coverage begins at +12° declination and progresses northward during the lifetime of the survey. The boundary between hemispheres is located at $\delta = +12°$ because weather statistics suggest that sky coverage will be more efficient from the Southern Hemisphere.

The fundamental unit of 2MASS coverage is the 6° by 8.5′ tile. Approximately 60,000 tiles cover the entire sky. Scheduling software at the telescope automates and optimizes the selection of tiles. Tiles must be within 0.2 airmasses of the transit airmass to be candidates for observation. Priority is assigned on the basis of available observing time in the future which largely dictates the strategy of progressing from equator to pole. The Northern hemisphere survey began by mapping the +12-18° declination band. When no more tiles were available at a given time of night the survey scheduler began covering the +18-24° declination band. Figure 1 reflects the coverage obtained in June through October. The Northern Hemisphere facility does not operate in August due to the Arizona summer monsoon. The coverage map represents all observations obtained to date. Some fraction (probably 30 to 50%) of these observations will be scheduled for repeat coverage

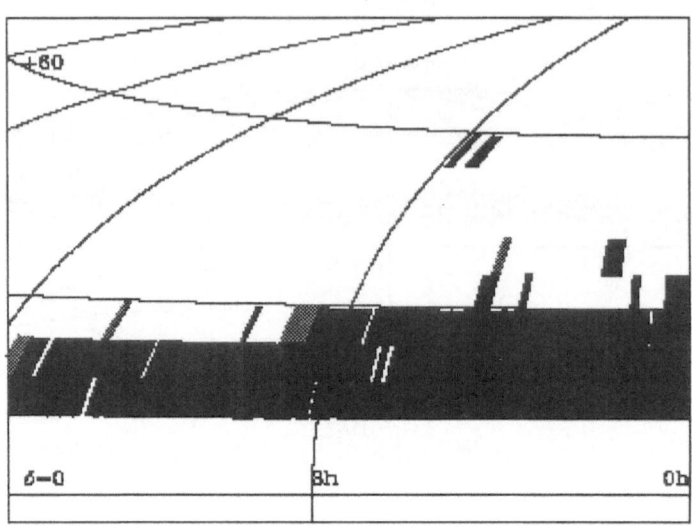

Figure 2. A magnified view of one equatorial quadrant showing detail at the level of individual 2MASS tiles.

because of poor photometric, airglow, or seeing conditions.

3. Calibration Strategy

2MASS calibration tiles are observed at one hour intervals throughout the night. Calibration observations begin and end the night. Calibration data is obtained in the same manner as regular survey data with the exception that calibration tiles are 1 degree in length rather than 6 degrees long. Each calibration tile is repeatedly observed 6 times during a calibration observation. Since each calibration tile yields 6 apparitions of the star, a single calibration series produces 36 apparitions of all of the stars in the 1-degree by 8.5-arcminute region. The majority of the calibration tiles contain a classical calibration standard, most of which were measured for the NIC-MOS standards program. Ultimately 2MASS will to internally calibrate the survey using all of the hundreds of stars which appear in each of these calibration strips.

Adjacent 2MASS survey tiles overlap their neighbors by 10% in right ascension. The survey scheduling software makes every effort to obtain tiles in contiguous blocks so that the comparison of photometry in the overlapping regions provides an additional test of photometric quality. At high galactic latitudes the number of stars in this overlap region will be sufficient to photometrically register adjacent strips to 0.01 magnitude. Global reprocessing of the survey data at the completion of operations will include an effort to internally photometrically register all of the survey data using these redundant observations.

4. Data Products

IPAC will begin production processing of the survey data taken to date (1.5 Terabytes at the time of this writing) in early 1998. The 2MASS project aims to distribute the first incremental release of survey data in mid-1999. The exact format and and volume of this release are still under discussion, but the project will aim to make these preliminary data products – point and extended source catalogs and images – largely resemble the final release products. Given the scheduled survey coverage, this first release should largely come from the declination range +12-24° in a band which completely circles the sky and thus covers a full range of galactic and extragalactic environments. The release may include some southern hemisphere data depending on the timing of the commissioning of that facility. The first sky coverage in the "southern" hemisphere will be the +6-12° declination band.

References

Milligan, S., Cranton, Brian W., Skrutskie, Michael F. 1996, "Development of a three-channel SWIR camera, for ground-based astronomical imaging", Proceedings of the SPIE, **2863**, pp 2-13.

Skrutskie, M.F., *et al.* 1997, in "The Impact of Large Scale Near-IR Sky Surveys," F. Garzon et al. (eds.), Kluwer (Netherlands), pp 25-32.

MIDCOURSE SPACE EXPERIMENT SURVEYS

STEPHAN D. PRICE, MICHAEL P. EGAN AND RUSSELL F. SHIPMAN
Air Force Phillips Laboratory
Geophysics Directorate/GPOB
29 Randolph Rd.
Hanscom AFB, MA 01731-3010
USA

MARTIN COHEN
Radio Astronomy Laboratory
University of California at Berkeley
Berkeley, CA 94720
USA

AND

MEHRDAD MOSHIR
IPAC
California Institute of Technology
MS 100-22
Pasadena, CA 91125
USA

Abstract. The Midcourse Space Experiment (MSX) conducted several infrared surveys during its 10 month mission which began in April 1996. We report on the progress in the analysis of these survey observations with an initial estimate of the instrument performance parameters. Initial results indicate that the surveys are at least as sensitive as the Infrared Astronomy Satellite (IRAS) survey, and have higher spatial resolution.

1. Introduction

The Midcourse Space Experiment (MSX) surveyed the entire Galactic Plane to a latitude of ±5°. In addition to the survey scans, raster scan observations toward selected directions probed Galactic structure more deeply.

N. Epchtein (ed.),
The Impact of Near-Infrared Sky Surveys on Galactic and Extragalactic Astronomy, 17-27.
© 1998 *Kluwer Academic Publishers.*

Band	Isophotal λ (μm)	λ (μm) 50% peak Intensity	Isophotal BW Δλ(μm)	0 mag flux (Jy)	Absolute Phot. Accuracy	Point Source Survey Sensitivity (Jy)
A	8.28	6.8 - 10.8	3.36	58.55	8%	0.1 - 0.2 - 0.4
B_1	4.29	4.22 - 4.36	0.104	194.5	12%	2.0 - 4.0 - 8.0
B_2	4.25	4.24 - 4.45	0.179	188.8	13%	2.5 - 4.0 - 8.0
C	12.13	11.1 - 13.2	1.72	26.51	5%	0.15 - 0.3 - 0.6
D	14.65	13.5 - 15.9	2.23	18.29	5%	0.1 - 0.2 - 0.4
E	21.41	18.2 - 25.1	6.34	8.75	13%	0.4 - 0.7 - 1.2

TABLE 1. SPIRIT III Spectral Bands

MSX also surveyed the sky not covered by IRAS both the inertial gaps and the zodiacal background.

1.1. INSTRUMENT

The infrared telescope on MSX is an off-axis, aspheric re-imaging optical system with an unobscured primary aperture of about 33 cm diameter. The five line-scanned focal plane arrays each consist of 8 in-scan columns and 192 cross-scan rows of detectors (pixels). (Band B has two slightly different narrow band filters, one spanning the upper half of the array, the other the lower.) Each detector is approximately square and is about 18.3" in extent. Four of the columns are offset in cross-scan with respect to the other four, providing critical (Nyquist) sampling in cross-scan for a single observation. To reduce telemetry requirements only half the columns are active. (All eight columns in band A, two on either side of the stagger in bands C and D, and two staggered columns in bands B and D.) Table 1 lists the band designation, the wavelength at half peak response, the isophotal wavelengths and bandwidths. The source function in the isophotal calculations is the Kurucz model Vega spectrum in Cohen et al. (1992); MSX used the Vega and Sirius spectrum in this article plus the calibrated spectra of the six additional standards published by Cohen et al. (1995) for in-flight calibration.

The MSX photometric bands have several major astronomically relevant properties. Bands B and D are centered on the 4.2 and 15 μm CO_2 atmospheric features; besides MSX, only the ISO camera is currently capable of doing photometry at these wavelengths. Band A is the most sensitive (having all 8 columns active) and covers a spectral region not previously well surveyed. Band C is a narrower analog of the IRAS 12 μm filter and the COBE/DIRBE band 5. Band E is a good analog of the COBE/DIRBE band 6, which is commonly taken as a comparison with the IRAS 25 μm band.

Owing to the diverse experimental objectives, MSX operations bear little resemblance to typical space-based astronomy missions such as IRAS or ISO. For a variety of reasons (see Mill et al., 1994) data acquisition was limited for infrared observations to 2-2.5 hours a day for a duty cycle of about 10%. Other spacecraft constraints limited continuous infrared observations to 36 minutes or less. Furthermore, observations were routinely made against the hard Earth. Thus, the astronomy observations were subject to the tail end of start up transients and small temperature variations from the thermal pulses caused by the Earth looking observations. These short term thermal variations were superimposed on a systematic increase in the operational temperature of the system as the solid hydrogen in the cryostat sublimed away from the thermal straps.

In spite of the dynamic state of the telescope, the uncertainties quoted in Table 1 are conservative. Three independent means of calibrating the instrument were used: a ground calibration in a vacuum chamber with sources which were well quantified and traceable to NIST standards; stellar standards; and well characterized black emissive reference spheres ejected from the MSX spacecraft. More than half of the on-orbit observations were in support of calibration and performance assessment experiments used to quantify the photometry as a function of instrument parameters. The conservative absolute photometric accuracies listed in Table 1 contain a few percent bias between calibrations and the estimated error in transferring the high signal to noise calibrations throughout the dynamic range of the sensor. The latter dominates the uncertainty. These uncertainties will be reduced within the next year by a global calibration which cross-ties the three methods and extends the calibration over the entire dynamic range. Relative accuracy, or the agreement of repeated observations on a source, is 2-3%. This is consistent with the fact the response of the pixels are uniform within 5% across an array.

1.2. EXPERIMENTS

The basic surveys were composed of scans between 130° and 180° in length, nominally at a scan rate of 0.125 deg/sec. The combination of scan rate, pixel size and the 72 sample/sec telemetry rate yields 2.8 samples/dwell. The scan rate was decreased somewhat toward the end of the cryogen life to partially compensate for the increase in dark current noise as the focal plane temperatures rose. In addition to the long scans, fifteen raster scans were made at various locations in the Galactic plane in the inner Galaxy and one in the anticenter direction.

The scan pattern was tailored for the objectives of each of the surveys. The Galactic plane was covered by 75 scans in longitude at constant latitude. The first survey began at the Galactic plane with subsequent scans offset by increments of 0.45° in latitude. This pattern produces a single survey in each of the B bands and redundant coverage in remaining bands. A second set of scans was interleaved with the first survey to provide four pass redundancy. There was inadequate time to complete the overlapping survey scans so quadruple coverage in the first and fourth quadrants is missing two of scans at > +4° latitude and extends to |b| < 3° in the second and third quadrants. Similarly, four pass redundancy was obtained by 64 scans over the two IRAS gaps. The geometry used to survey the IRAS gaps was that of a defined rotation pole, a zenith angle and azimuth limits; the scans were roughly along ecliptic longitude. The 64 scans required to cover the gaps four times spanned a range of sun centered longitudes from 60° to 173° . To this are added five ecliptic pole to pole scans spanning sun centered longitudes 25° to 30°, providing global observations of the zodiacal emission.

The raster scans of selected regions along the Galactic plane consisted of 21-25 legs, each spanning approximately 3° in latitude, centered at constant longitude, with a scan rate of 0.05 deg/sec. Successive raster legs were offset by a quarter to a third of a pixel. The longer dwell time and redundancy increases the sensitivity by about a factor of about eight compares to the survey scans. Alternatively, one could take advantage of the oversampling and use image enhancement techniques to resolve sources in those regions where even MSX is confused. The observations from the deep raster scans will be used to provide truth tables to assess completeness and reliability of the larger area survey.

1.3. DATA PROCESSING

One of the primary data products from the survey will be the point source catalog. For point source identification and parameter estimation, a two dimensional source extraction algorithm is used because the point response functions have about the same support in both cross-scan and in-scan. Since dark current subtraction errors and other sources of pattern noise are correlated in-scan more than cross-scan, a form of pseudo-median estimator was used to filter the data stream for each pixel. An average of the cascaded MINIMAX and MAXIMIN operators, $\frac{1}{2}[MINIMAX(MAXIMIN\{S_L\})+ MAXIMIN(MINIMAX\{S_L\})]$ where S_L is the data in the window around the current point (see Pratt 1991), separates point sources from the background. This filter was chosen because it not only eliminates the striping from dark current subtraction residuals and other pattern noise in the data but also because it has superior characteristics in crowded regions and at inflection in the data. As with all (pseudo-) median filters, low amplitude (SNR $\leq$ 3) sources are biased by a small amount (5-8%) to lower values.

Two data streams resulting from the filtering (low frequency background, and high frequency, containing point sources). The background data is averaged into 0.25° square pixels and written to a file which provides the basis for analyzing the zodiacal background and the emission from interstellar dust along the galactic plane without the bias of point sources. The high frequency data is processed to extract the point sources .

The point source data is first convolved with a model of the point response function (PRF), a B_3 or cubic B-spline with the same full width at half maximum (FWHM) as the measured PRF. This model is used rather than the position dependent PRF to increase computational speed for the search step. The convolution almost realizes the visibility gain of matched tuned filtering of the data with the system response to a point source . The visibility is the co-add increase in signal to noise ratio (SNR) over that from the single sample. Dead and anomalous pixels (<5% of the total of 3840 pixels) are removed as a first step which make the visibility a weak function of cross-scan position. Potential sources have amplitudes of 4.5 times the noise after filtering, which equates to a SNR of 2.7 for the final extracted point sources, which do not have the matched-filter gain. The flux and position of each potential source are determined using a three parameter simultaneous χ^2 minimization of the data and a point source model, as follows:

$$\chi^2 = \sum_{i=1}^{N} \left[\frac{\rho_i - \sum_{k=1}^{M} f_k H(x_i - \xi_k, y_i - \eta_k)}{\sigma_i} \right]^2 .$$

Here ρ_i is the data value at point i while (x_i, y_i) is the detector position and σ_i^2 is the variance of the noise in the data (detector and Poisson). The data model consists of M point sources of flux and position f_k, (ξ_k, η_k) convolved with the PRF, H. The PRF is band dependent and has been determined on orbit. (The initial results presented in the next section use an averaged PRF for each band. The next step is to include the position dependent PRFs.).

2. Results

2.1. POINT SOURCES

This section constitutes a progress report which provides the realistic estimates of the survey sensitivities which appear in Table 1. Figures 1 through

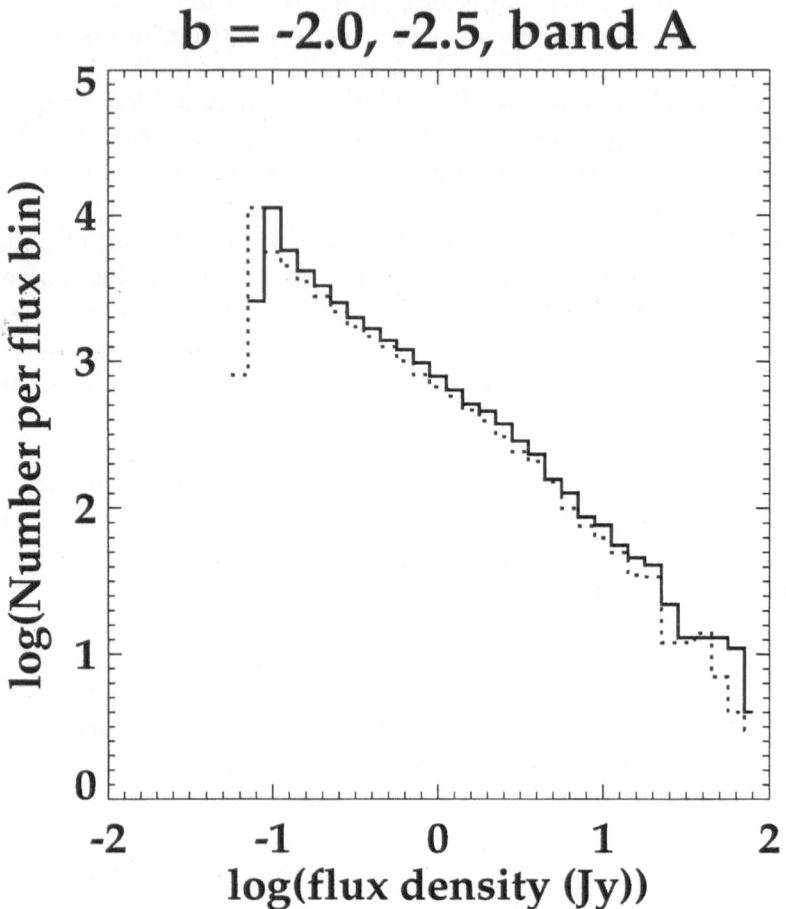

Figure 1. dLog N vs log F histogram of point sources extracted from MSX band A data. The solid line plots the data for the 180° long scan centered on b = -2.0°, while the dotted line shows results for the scan centered at b = -2.5°. The flux cutoffs are 96 and 80 mJy, respectively.

5 show the dLog N vs. Log F plots in dLog F increments of 0.1 for each of the radiometer bands in two first-fourth quadrant scans at b = -2.00° and -2.5°. These scans form the basis of our early analysis and estimation of the survey performance. At these latitudes, the complex background from interstellar dust is considerably reduced compared to the scan at b = 0. Also, these scans avoid the highest source density regions so an estimate of the performance can be obtained from the source distribution plots without the added complication of confusion.

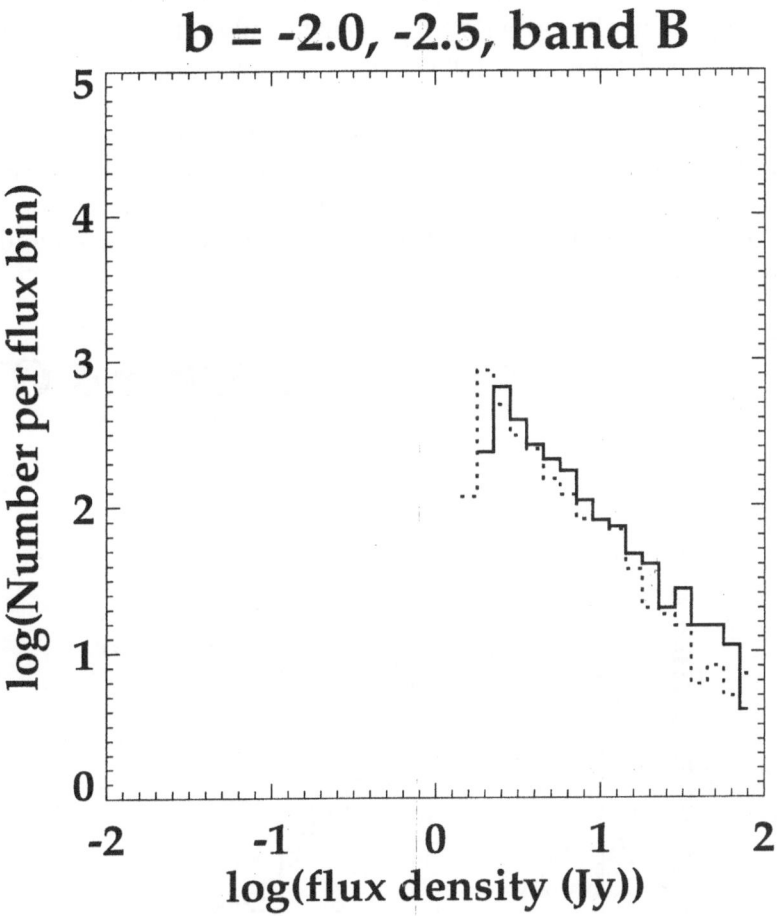

Figure 2. dLog N vs log F histogram of point sources extracted from MSX band B data. The solid line plots the data for the 180° long scan centered on b = -2.0°, while the dotted line shows results for the scan centered at b = -2.5°. The flux cutoffs are 2.0 and 2.4 Jy, respectively.

The turnover in each Figure is defined by the SNR criterion. However, spurious sources evidently increase the number of objects in the last one or two flux bins well marking the statistical limit for real sources. Note, that the flux at the limits are underestimated by 5-8% as the background filter bias has not been removed in these plots. Anomalous pixels have been removed from the process which by definition includes those with response is greater than 5% of the mean. Therefore, the SNR cutoff is a good representative for an absolute cutoff in flux. Note that in all cases the cutoff for the later scan (taken 5 September 1996) is at about 20% higher flux than the earlier scan (taken 19 July). This reflects the increase in dark current noise as the focal plane temperature gradually rose as the solid hydrogen cryogen sublimed away. The first and second flux limits for extracted sources in Table 1 are representative for the beginning and end of the first and fourth Galactic quadrant survey and for the IRAS gap at ~342° ecliptic longitude. The highest flux limit in Table 1 is for the highest noise levels observed during the surveys of anti-center region and the IRAS gap at ecliptic longitude ~162° .

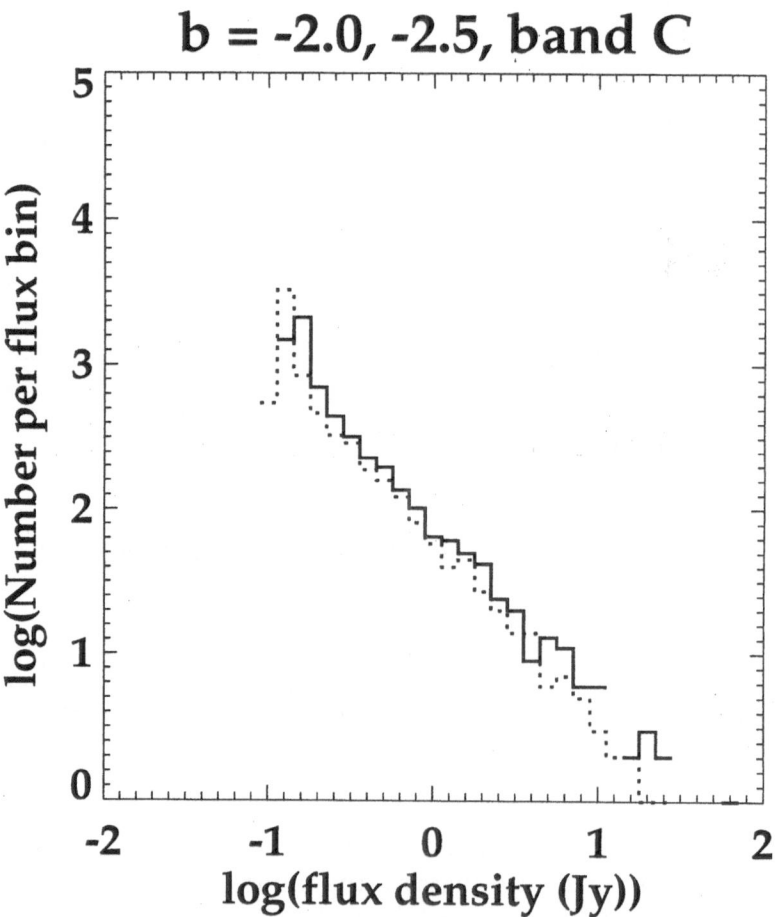

Figure 3.   dLog N vs log F histogram of point sources extracted from MSX band C data. The solid line plots the data for the 180° long scan centered on b = -2.0°, while the dotted line shows results for the scan centered at b = -2.5°. The flux cutoffs are 150 and 120 mJy, respectively.

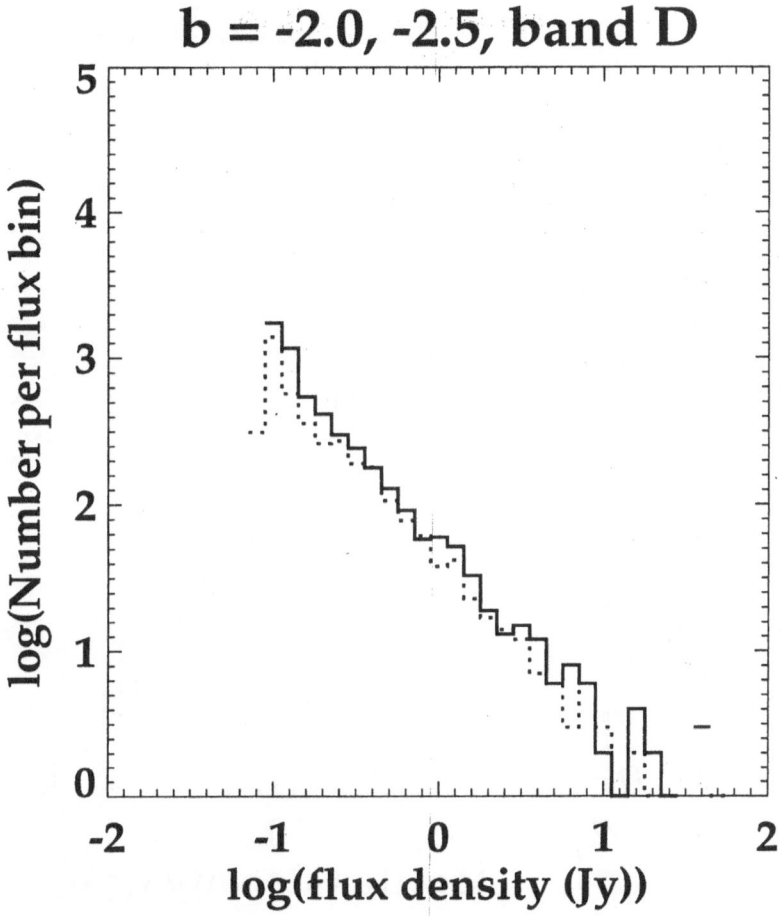

Figure 4. dLog N vs log F histogram of point sources extracted from MSX band D data. The solid line plots the data for the 180° long scan centered on b = -2.0°, while the dotted line shows results for the scan centered at b = -2.5°. The flux cutoffs are 100 and 110 mJy, respectively.

The MSX survey of the inner Galaxy is about as sensitive in Bands C and E as the IRAS instrumental limits at 12 and 25 μm , Band A is about 4 times more sensitive. Thus, one of the original goals of the astronomical experiments on MSX will be realized: to complete the census of the mid-infrared sky to the IRAS flux limits. The MSX point response is about 30 times smaller than those in the IRAS 12 and 25 μm bands. Consequently, MSX can probe 3 to 5.5 times more deeply into the IRAS confused areas.

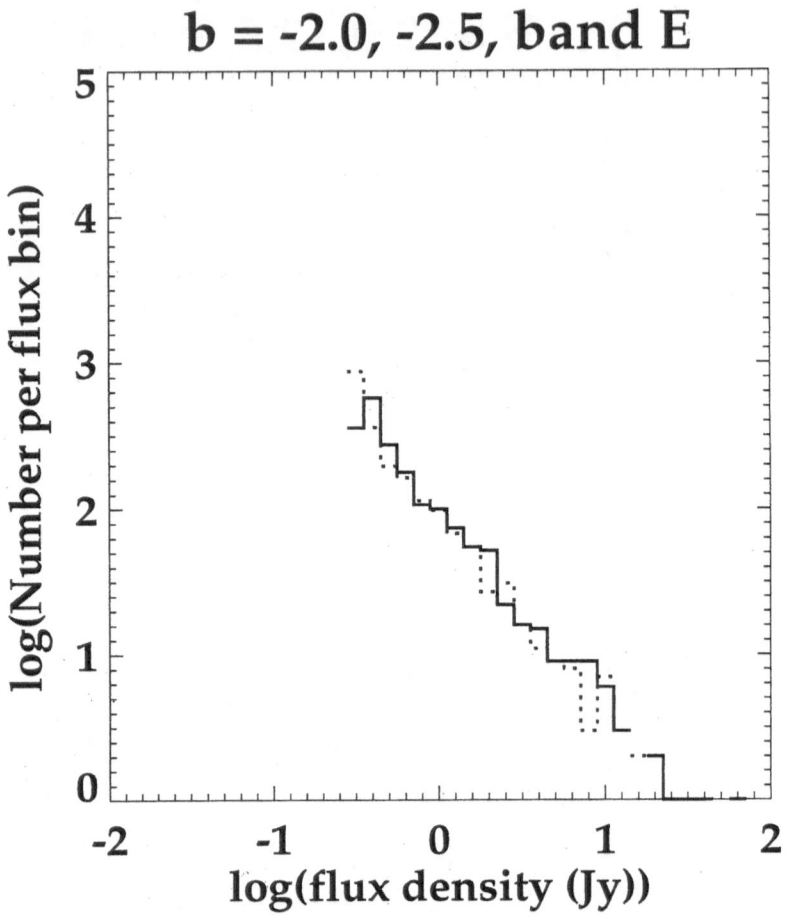

Figure 5. dLog N vs log F histogram of point sources extracted from MSX band E data. The solid line plots the data for the 180° long scan centered on b = -2.0°, while the dotted line shows results for the scan centered at b = -2.5°. The flux cutoffs are 370 and 320 mJy, respectively.

2.2. IMAGE DATA

In addition to the point source data, image data is produced for every scan. The increased resolution of MSX data over the IRAS data shows that there is highly structured IR emission throughout the galactic plane. Most strik-ing is the complexity of the interstellar dust emission in Band A above a radiance threshold of several MJy/sr. The background in this band is undoubtedly enhanced by the strong 6.6, 7.7 and 8.3 micron PAH/HAC emission features associated with interstellar dust. In addition to the structured emission, we have found over two thousand dark clouds silhouetted against this bright background within half a degree of the plane (the latitude limits we have searched to date). The clouds are sharp edged, dark (implying large optical depth) and cold ($\leq 10K$ as deduced from the absence of emission in the IRAS bands). The extinction edges of the clouds do not change across the MSX wavebands, which means that the extinction properties of the dust do not change between 7 to 25 μm. This implies that the dust grains are quite large, which is consistent with a very low temperature and grain mantle growth due to freeze-out of molecules onto grains. The clouds appear to be a new population of very cold, dense, molecular cores.

3. Conclusions

The MSX survey data will serve to complete the census of the mid-infrared sky begun by the IRAS satellite. In addition to completing observations in the IRAS gaps, the enhanced resolution of the MSX infrared instrument allows accurate cataloging of stars in the Galactic plane, where the IRAS data products suffered from confusion due to high source density and large detector size. Viewed as images, the MSX data has shown complex extended emission structures along the plane, and reveals a population of cold, dense, cores in the interstellar medium.

References

Cohen, M., Walker, R.G. , Barlow, M.J. and Deacon, J.R. (1992) Spectral Irradiance Calibration in the Infrared. I. Ground-Based and IRAS Broadband Calibration, *Astron. J.*, **Vol. 104**, pp. 1650-1657

Cohen, M., Witteborn, F.C., Walker, R.G., Bregman, J.D., and Wooden, D.H. (1995) Spectral Irradiance Calibration in the Infarared. IV. 1.2-35 μm Spectra of Six Standard Stars, *Astron. J.*, **Vol. 110**, pp. 275-289

Mill, J.D., O Neil, R.R., Price, S., Romick, G.J. , Uy, O.M., Gaposchkin, E.M., Light, G.C. , Moore Jr., W.W. , Murdock, T.M., and Stair jr., A.T. (1994) Midcourse Space Experiment: Introduction to the Spacecraft, Instruments and Scientific Objectives, *J. Spacecraft and Rockets*, **Vol. 31**, pp. 900-907

Pratt, W.K. (1991) *Digital Image Processing*. John Wiley and Sons, New York.

II- Galactic Structure

THE IMPACT OF IR SURVEYS ON GALACTIC STRUCTURE

O. BIENAYMÉ
Observatoire de Strasbourg,
11 rue de l'Université, F-67000 FRANCE

Abstract. Most of the galactic stellar mass is in the galactic disc and IR observations allow one to observe through nearly the entire disc, giving access in detail to its morphology, from the inner bar-bulge, the arms and inner ring, the outer limit of the exponential discs to the dependence of the scale lengths with stellar populations, while the brightest halo giants completely probe the stellar halo morphology and structure.

Kinematic observations can be performed to complete IR multicolour star counts. Such complementary data will strongly constrain the morphology of the stellar components and give clues to the scenario of galactic formation and evolution. Some of the expected results concerning the internal dynamics of our Galaxy are constraints on the galactic potential in the halo, a direct measure of the rotation of the inner bar, kinematic gradients in the stellar discs and dependence with age.

The local stellar Luminosity Function can be much better determined using proper-motions, separating discs and halo Luminosity Functions. It will also be the only means to accurately identify existing intermediate populations like the thick disc, as well as other hypothetical populations (extended bulge or accreted satellites).

1. Galactic components and Parameters of the Milky-Way

The main constituents of the Milky Way galaxy are the stellar populations, the ISM and the dark matter. The stellar component is usually separated into three populations with different morphology, kinematics and also ages and abundances; evolutionary links between these quantities are far from fully understood. Stellar populations are the thin and thick discs, with a total mass $M_d \sim 4~10^{10}~M_\odot$, rapidly rotating with low internal dispersion, vertical dispersions in the range of $5\,\mathrm{km\,s}^{-1}$ to $18\,\mathrm{km\,s}^{-1}$ for the thin disc

31

N. Epchtein (ed.),
The Impact of Near-Infrared Sky Surveys on Galactic and Extragalactic Astronomy, 31-40.
© 1998 *Kluwer Academic Publishers.*

(Gomez et al, 1997) and $40\,\mathrm{km\,s^{-1}}$ for the thick disc. The bulge population is in a slightly flattened spheroid of mass $M_b \sim 2\ 10^{10}\ M_\odot$, rapidly rotating ($\sim 100\,\mathrm{km\,s^{-1}}$) and with a high internal rotation ($\sim 100\,\mathrm{km\,s^{-1}}$). The Metal-Poor Halo has a mass $M_h \sim 10^9\ M_\odot$ and low rotation, its contribution to the dynamical mass of the Galaxy is negligible. Finally, at large radii the main contributor to the mass is the dark matter component: at large R the mass is proportional to R, $M_{D.M}(R) \sim V_c^2 \sim 10^{10}\ R$ if the rotation curve ($=V_c$) remains flat.

Thus the stellar mass is essentially in the discs and mainly in the lowest mass stars. Most of the stellar mass is then emitting in IR and in the galactic plane, strongly obscured in visible band. As a consequence, IR observations are necessary for both reasons that low mass stars are bright in the IR and that in IR bands the extinction by interstellar matter is minimal.

Therefore DENIS, 2MASS, ISOGAL or MSX are the most promising surveys for the development of new and global views on the structure of our Galaxy (and of external galaxies).

2. Photometry and galactic morphology

2.1. SURFACE PHOTOMETRY

To study the galactic structure, wide field surface brightness surveys were obtained during the Pioneer and, more recently, COBE/DIRBE experiments. The line-of-sight integral of luminosity distribution maps are interpreted using models and fitting free parameters. The lowest galactic latitude fields (less than 20°) are affected by interstellar obscuration, while at higher latitudes interstellar extinction is less serious. Infrared observations reduce the difficulties of extinction. The recent DIRBE observations have been investigated by numerous groups leading to detailed parameterization of the Galaxy: however some of the structural properties deduced from surface photometry data alone are fully model dependent, and it can be shown that very different 3D models can produce nearly identical surface densities on the sky (Zhao, 1997).

2.2. IR COUNTS

IRAS sources gave a deep and large-scale view of the Galaxy. Sources are dominated by late type giants AGB and dust shell stars; they form two (thin and thick) discs and the very identifiable bulge. From IRAS, Garwood and Jones (1987) have shown that the halo is consistent with a $R^{1/4}$ law and with the local density of Population II subdwarfs.

Much better resolution of IR counts from DENIS and 2MASS will provide wide digital sky surveys with reliable photometric calibrations and uniform acquisition and reduction (visible bands photometric surveys from Schmidt plates are under progress, however the final photometric accuracy will not be so accurate).

Counts compared to surface photometry carry information on the density distribution along the line-of-sight, and the distances can be statistically determined as long as the luminosity function is effectively known everywhere. In practice, galactic structure parameters are found through models with building blocks having the relevant astrophysical ingredients. The necessary ingredients for such models are: parametric density laws for the different stellar populations; accurate Luminosity Functions that are now achievable from Hipparcos or similar data; calibration of the colour or magnitude dependence with metallicity; modeling of the absorption either by some regular absorption law or by ad hoc structures to mimic clouds or arms.

Comparison of models to stellar counts allows one to determine the morphology of components. While counts in three colours are necessary to accurately discriminate the effect of metallicity from morphology, extinction must be calibrated with more colours. For instance at least IJH and K colours should be used.

Converging results are now obtained; for example, from the TMGS (Hammersley et al, 1995), the Sun is at $15.5 \pm 3 \, \text{pc}$ above the plane and there is a possible tilt of the local old disc with respect to the average HI or stellar disc of the inner Galaxy. Stanek et al (1997), using red clump stars towards the bulge, determine the inner bar orientation. Ruphy et al (1996), with DENIS data, find a scale length for the disc, $2.3 \pm 0.1 \, \text{kpc}$, and the cutoff radius of stellar discs at $R = 15 \pm 2 \, \text{kpc}$; and Ruphy (1996) obtains results concerning the morphology of inner arms and ring. Other determinations concerning the disc, thick disc scale height and halo should be obtained.

3. Kinematics and dynamics

The analysis of stellar kinematics and Galactic potential is linked to the study of the spatial distribution of stars in the Galaxy, since they are related through the Boltzmann and Poisson equations. Measuring all the visible density and mass distribution from general star counts and the gas cloud density gives only a small fraction of the total amount of the dynamical mass that is deduced from the kinematics of the galactic constituents. As in many spiral galaxies, most of the Galactic mass is unseen and unknown.

We could classify the main potential, or mass, "tracers" as 1) those related to the determination of the rotation curve: HI and CO clouds, young disc populations like cepheids or young open clusters... 2) those related to the study of the density distribution perpendicular to the galactic plane and the determination of the plane galactic density: gas flaring and stellar samples up to 1 or 2 kpc towards the Galactic poles... 3) samples related to a 3D halo analysis, such as RR Lyrae or BHB stars, globular clusters and at larger distances galactic satellites (see reviews by Fich & Tremaine, 1991 and Crézé, 1991).

It is important to find tracers corresponding to very different scales in order to obtain an accurate determination of the potential at each scale. This will help to clarify some unsolved questions: what is the exact mass density at large galactic radius; what is the shape of the dark matter distribution -spherical dark halo or flat-; is it triaxial; is there a stellar population with intermediate angular momentum? In particular, many observational efforts linking deep star counts to radial velocities and proper motions measurements are still needed.

Another essential question is the determination of the evolutionary relationship between stellar populations in the different components of the Galaxy. Links between kinematics, stellar ages and chemical abundances allow one to build the dynamical history of the Galaxy from the old, metalpoor stars of the stellar halo to the present, young disc stars (Gilmore et al, 1989). Such studies provide the opportunity to understand the formation, evolution and structure of galaxies in general and to discriminate between formation scenarios: 1) Halo stars formed during a free-fall collapse of the proto-Galaxy and disc stars formed later with enriched gas in a rapidly rotating disc (Eggen et al, 1962); 2) An alternative model is that the halo formed as a merger of dwarf galaxies over several Gyr (Searle & Zinn, 1978). Reality looks certainly much more like a combination of these two scenarios.

It is more fruitful to combine kinematics and morphological analysis of stellar populations that are strongly correlated in the case of nearly steady state systems. Solutions of dynamical systems given by the Boltzmann equation allow one to quantify the relations between the density distribution and the velocity distribution. For instance, explicit solutions are known in the case of the determination of the force perpendicular to the galactic plane. A good approximation is a one dimensional problem and the general solution for an isothermal population (constant velocity dispersion σ) is known: the vertical density is $\rho(z) = \rho(0)\ exp(-\Phi(z)/\sigma^2)$ where $\Phi(z)$ is the potential. More generally, the virial theorem relates the velocities, the scale lengths and the potential (through the mass M): for a rotating disc, $M \sim V^2/R^2$, or for a non-rotating population, $M \sim \sigma^2/R^2$.

In conclusion, it is more constraining to analyse simultaneously galactic morphology and kinematics. A straightforward example is the kinematical determination of the scale length $(2.5 \pm^{0.8}_{0.6}$ kpc) of the Galaxy that can be deduced from a local sample of stars with known velocities (Fux & Martinet,

1994). Important results are obtained in the study of the galactic bulge-inner bar (see below) using the kinematics to discriminate between models that were compatible using star counts or surface photometry alone. An essential by product is the measure of the potential and of total (dynamical) mass.

Concerning the galactic potential, it can be recovered from the rotation curve assuming the shape of the mass distribution (spherical, flat...). The rotation curve at a given radius constrains mainly the amount of mass inside this radius. Forces from dark matter account for half of this mass at the solar radius. This estimate depends slightly on the mass assigned to the stellar disc and its scale length.

3.1. NEED FOR A LARGE KINEMATICS SAMPLE

The second important aspect of kinematic study is the need for a large sample of stars with radial velocities or, more easy to obtain, with proper motions. The necessity for large set of data is closely related to the (numerically unstable) inverse Laplace transform at least implicitly faced in dynamical analysis. A simple example is the splitting of a velocity distribution into a summation of various gaussians: very different solutions may coexist having the same statistical meaning unless a sufficient quantity of data is available. Large surveys will soon be available by combining proper motion surveys like PMM or others to IR survey. They would help to clarify the identification of kinematic populations.

4. Galactic components

4.1. BAR-BULGE

From a qualitative comparison of the optical and radio properties of our Galaxy with spirals, de Vaucouleurs & Pence (1978) classified it as an SAB(rs) with an inner bar (see their fig. 6). A tilted bar was proposed to explain the kinematics of the HI and CO at the Galactic center. High latitude parts of the bulge are accessible in visible bands, but only IR surveys allow one to explore in more detail the inner structure of the Galaxy. The bar was detected in the 2.4-micron observations of the Galactic center of Matsumoto et al (1982).

A large amount of work using COBE/DIRBE data confirmed an inner bar (Dwek et al, 1995, see also a recent review by Weinberg, 1996) that is tilted relative to the line-of-sight, and also the presence of arms or an inner ring (Binney et al, 1997). These results are strongly model dependent and it has been shown (Zhao, 1997) that the differences between the inner bar properties obtained by various authors is just due to the non-unique deprojection of the Galactic bar. A bar with various orientations and axis ratios would produce the same projected surface brightness. Stanek et al (1997) identified red clump stars from V, I magnitudes in 12 fields obtained by the OGLE collaboration and presented a more accurate description of the density along the line of sight constraining more tightly the inner structures (bar oriented at $20° - 30°$ to the line of sight and axis ratio 3.5:1.5:1).

On the other hand, kinematic data give independent measurements and radial velocities (Tiede & Terndrup, 1997), or proper motions from Spaenhauer et al (1992), used by Zhao et al (1996), allow one to determine the bar properties. Anisotropy of proper motions are compatible with an axis ratio c/a of 0.7, while the rotational velocity and tangential dispersion where deduced from the radial velocity sample. Both methods find a bar pattern speed of $70 - 77\,\mathrm{km\,s^{-1}}$/kpc (model dependent), while we remark that a really direct determination of the pattern speed can be obtained with accurate (1 mas/y) proper motions measured at the extremity of the bar, since we just have to wait and "see" the bar rotating.

The most powerful analyses are a combination of available information, photometric and kinematic, in a unique self-consistent dynamical description to take full advantage of correlations. It will give the full 3D mass and potential distribution and define the part of random to periodic or regular stellar motions.

Finally, microlensing detections towards the bulge can certainly give the bulge stellar mass function and consequently the total stellar mass in the bulge.

4.2. STELLAR DISCS

According to external spiral galaxies, stellar discs are exponential. The Galactic thin disc is a continuum of discs with varying age, metallicity, kinematics and morphology. Scale-heights range from 75 pc to 250 pc, and vertical velocity dispersions from 5 to $20\,\mathrm{km\,s^{-1}}$. Most recent optical scale length determinations favour short values ~ 2.5 kpc (Robin et al, 1992).

It is very difficult to find evidence from star counts in the galactic plane for ellipticity of the stellar disc at the solar galactic radius, particularly if we are on one of the symmetry axes. Signatures from kinematics give more reliable constraints (Kuijken & Tremaine, 1994): Oort's constants C and K, the LSR radial velocity, show no evidence for non-axisymmetry, while recent analysis (Gomez et al, 1997) from Hipparcos data favours a null vertex deviation (velocity ellipsoid pointing towards the galactic centre). Likewise the axis ratio σ_u/σ_v is observed to be about ~ 0.5 and can be a signature of ellipticity (Kuijken & Tremaine, 1994), but may also simply be explained if the rotation curve is slightly decreasing locally, or if the kinematic and density scale lengths are very different (Bienaymé & Séchaud, 1997).

A global analysis of the stellar discs is possible with OH/IR stars and Planetary Nebulae that can be identified near the plane at large distances and for which radio observations supply accurate radial velocities. Such data give a unique opportunity to measure stellar density, orbital structure and kinematic gradients over large areas of the Galactic disc (te Lintel Hekkert, 1990; Sevenster et al, 1995; Durand et al, 1996). Systematic searches closer to the Galactic plane are possible (Kistiakowsky, 1995) and are promising for the study of younger populations and kinematic-gradient.

The identification of OH/IR stars has largely been done with IRAS and improvements of the detection and identification very close to the plane is expected with new IR surveys, while systematic searchs in radio (OH) are done in the bulge region (Sevenster et al, 1997).

OH/IR stars with known radial velocities can probe the disc dynamics if the sample is complete, and without determining the distance of individual stars. Bulge-to-disc links and their extensions are thus kinematically accurately defined. Kinematical scale lengths of disc and thick disc were measured. More extended surveys will allow one to define the relation between kinematical scale lengths and ages to understand the dynamical heating, both with age and galactic radius.

4.3. K_Z FORCE PERPENDICULAR TO THE GALACTIC PLANE

Comparing scale heights of stellar discs to the vertical velocity dispersions σ_w of their constituents gives direct access to the vertical potential. In the academic case of a stellar sample distribution, $\rho_*(z)$, where σ_w does not vary with height and when distances are below 1 kpc, we have $\Phi(z) = \sigma_w^2 log(\rho_*(z)/\rho_*(0))$. Then, the local dynamical mass ρ_{dyn} may easily be deduced from the potential. There has been a long controversy as to whether the dynamical mass is in excess of the total "observed" mass density ($0.09\,M_\odot pc^{-3}$), the main discussion being concerned with the quality of used samples. A summary of the most recent results are given by Crézé (1991) and Kuijken (1995), where it appears that the most accurate samples indicate no need for a flat disc of dark matter. Though less constraining, vertical HI gas dynamics favour this conclusion (Boulares & Cox, 1990; Malhotra, 1995).

The most recent result is obtained with Hipparcos data providing the first, volume limited and absolute limited homogeneous tracer of stellar density and velocity distributions in the solar neighbourhood. Crézé et al (1997), from A-F dwarf samples, find the local dynamical density as $\rho_0 = 0.076\pm0.015\,M_\odot pc^{-3}$, a value well below all previous determinations leaving no room for any disc shaped component of dark matter.

4.4. THICK DISC

The thick disc (TD) is identified by star counts at high galactic latitude (Reid & Majewski, 1993) and also by distinctive kinematic signatures that are constant at various heights above the galactic plane (Soubiran, 1993). The TD has a large rotational velocity, looking more like the thin disc populations than those of the halo, but its kinematics remain fully separate from the disc. This is very probably the signature of a merging event in the early phase of the Galactic disc's formation (Robin et al, 1996; Ojha et al, 1996). This may explain also the absence of a vertical abundance-gradient (Gilmore et al, 1996).

The thick disc vertical velocity dispersion ($38 \, \mathrm{km \, s^{-1}}$, Beers et al, 1996) and scale height (750 pc, Robin et al, 1996; Ojha et al, 1996) can be used to estimate the K_z force at large z (1-2 kpc). We estimate that it is consistent with no massive disc of dark matter (the plane-parallel approximation is no longer valid so far from the plane and a complete 3D modeling would be necessary).

4.5. THE GALACTIC HALO

Understanding and modeling the 3-dimensional dynamics of halo stars has progressed rapidly during the last few years; see, for example, models like the Stäckel potential with three explicit integrals of motion or models with action integrals (Dehnen & Binney, 1996) or with orbit computations (Flynn et al, 1996).

The quantity of available kinematic data for distant halo stars is surprisingly low, and most of our knowledge is based on local kinematics.

Halo samples with 3-D kinematics seem to show substructures that could be fragments of destroyed globular clusters or else... (Majewski et al, 1996). In fact, the mixing time for halo stars is a few Giga-years and relaxation is not fully achieved (Tremaine, 1993). We expect partial mixing to leave traces of initial formation or accretion period. Non-stationarity will limit the description of the Galactic potential, but it is also a unique chance to find explicit and "still" living traces of Galactic formation.

4.6. LOCAL KINEMATICS

Proper motions can be obtained with an accuracy of 6-8 mas/year by reasonable means. Combined with photometric IR surveys, this is sufficient to differentiate distant giants and close dwarfs. Difference in absolute magnitudes between the reddest giants and dwarfs is large and consequently, at a given apparent magnitude, the difference in proper motions is large. Thus a detailed description of the local kinematics could be obtained from the kinematics of neighbouring dwarfs.

The local kinematics describe the continuous range of thin discs the thick disc and the halo. Possible links between these components and intermediate populations would be revealed by groups of stars with intermediate properties, velocities dispersions and velocity lags. To find unknown and faint new populations is certainly one important goal: the existence of a possible bulge-like (metal rich) population (Grenon, 1990) or (metal poor) accreted satellite (Preston et al, 1994) with rotation around $100\,\mathrm{km\,s^{-1}}$ at the solar galactic radius could be established more clearly.

Metal-poor Blue Main Sequence stars near the solar circle have been discovered by Preston et al (1994) as a new kinematic population. These stars have an isotropic velocity dispersion and relatively large mean rotational velocities of about $128\,\mathrm{km\,s^{-1}}$. Preston et al suggest they are probably accreted from dwarf spheroidal satellites. It will be essential to check and extend the detection of this population out of the galactic plane, since it will be the best potential tracer at intermediate distances around 3-6 kpc out of the Galactic plane.

The kinematics from a very large data base is certainly the only tool for identifying accurately such faint populations that could be the remnant of an intermediate, rapid and important event during the formation or evolution of the Galaxy. The practical analysis of such data will need to establish definite IR-LF for each population. Multicolour calibrations must first be obtained from close red stars over a large range of metallicity. Inversely, kinematic signatures can be used to distinguish stellar populations and to determine more accurately their LF. Analysis of kinematics and the LF cannot be separated, but the counter part is that such an analysis will provide a more definite and exact separation between populations.

References

Beers T.C., Sommer-Larsen J., 1996, *Ap.J. Suppl.* **96**, 175

Bienaymé O., Séchaud N., 1997, *A. & A.*, **323**, 781

Binney J., Gerhard O., Spergel D., 1997, *M. N. R. A. S.* **288**, 365

Boulares A., Cox D.P., 1990, *Ap. J.* **365**, 544

Crézé M., 1991, *The Interstellar Disk-Halo Connection in Galaxies*, IAU Symp. **144**, 313

Crézé M. et al, 1997, *A. & A.*, (in press) *astro-ph/9709022*

Dehnen W., Binney J., 1996, *Formation of the Galactic Halo .... Inside and Out*, ASP Conf. Series **92**, eds. H. Morrison & A. Sarajedini, 391

de Vaucouleurs G., Pence W.D., 1978, *A. J.* **83**, 1163

Durand S., Dejonghe H., Acker A., 1996, *A. & A.* **310**, 97

Dwek E., Arendt R.G., Hauser M.G. et al., 1995, *Ap. J.* **445**, 716

Eggen O.J., Lynden-Bell D., Sandage A., 1962, *Ap. J.* **136**, 748

Fich M., Tremaine S., 1991, *Ann. Rev. Astron. Astrophys.* **29**, 409

Flynn C., Sommer-Larsen J., Christensen P.R., 1996, *M. N. R. A. S.* **281**, 1027

Fux & Martinet, 1994, *A. & A.* **287**, L21

Garwood R., Jones T.J., 1987, *P. A. S. P.* **99**, 45
Gilmore G., Wyse R., Jones J.B., 1996, *A. J.* **109**, 1095
Gilmore G., Wyse R., Kuijken K., 1989, *Ann. Rev. Astron. Astrophys.* **27**, 555
Gomez A.E., Grenier S., Udry S., Haywood M., 1997, Hipparcos Venice'97 (ESA)
Grenon M., 1990, *ESO/CTIO Workshop on bulges of galaxies*, 150
Hammersley P.L., Garzon F., Mahoney T., Calbet X., 1995, *M. N. R. A. S.* **269**, 753
Kistiakowsky V., Helfand D., 1995, *A. J.* **110**, 2225
Kuijken K., 1995, *Stellar Populations*, IAU Symp. **164**, 198
Kuijken K., Tremaine S., 1994, *A. J.* **421**, 178
Majewski S.R., Munn J.A., Hawley S.L., 1996, *Ap. J.* **459**, L73
Malhotra S., 1995, *Ap. J.* **448**, 138
Matsumoto T. et al, 1982, in the Galactic Center ed. G. Riegler & Blandford (New York: AIP), 48
Ojha D. K., Bienaymé O., Robin A. C. et al, 1996, *A. & A.* **311**, 456
Preston G.W., Beers T., Shectman S.A., 1994, *A. J.* **108**, 538
Reid N., Majewski S.R., 1993, *Ap. J.* **409**, 635
Robin A.C., Crézé M., Mohan V., 1992, *A. & A.* **265**, 32
Robin A.C., Haywood M., Crézé M. et al, 1996, *A. & A.* **305**, 125
Ruphy S., 1996, *Thesis*, Université de Paris VI
Ruphy S. et al, 1996, *A. & A.* **313**, L21
Searle L., Zinn, 1978, *Ap. J.* **225**, 357
Sevenster M.N., Dejonghe H., Habing H.J., 1995, *A. & A.* **299**, 689
Sevenster M.N. et al, 1997, *A. & A. Suppl.* **122**, 79
Soubiran C., 1993, *A. & A.* **274**, 181
Spaenhauer A., Jones B.F., Whitford A.E., 1992, *A. J.* **103**, 297
Stanek et al, 1997, *Ap. J.* **477**, 163
te Lintel Hekkert P., 1990, *Ph D Thesis The evolution of OH/IR stars and their dynamical properties*, Leiden University
Tiede G.P., Terndrup D.M., 1997, *A. J.* **113**, 321
Tremaine S., 1993, *Back to the Galaxy*, AIP Conf. Proc. **278**, eds. S.S. Holt et al, 599
Weinberg M. D., 1996, *Barred galaxies* ASP Conf. Series **91**, ed. R. Buta et al, 516
Zhao H.S., 1997, *astroph9705046*
Zhao H.S., Rich R.M., Biello J., 1996, *Ap.J.* **470**, 506

RESOLVED DENIS AND UKIRT INFRARED STELLAR OBSERVATIONS TOWARDS THE GALACTIC CENTRE

M. UNAVANE, G. GILMORE

Institute of Astronomy,
Cambridge CB3 0HA, UK

Abstract.

IJK$_s$ DENIS images, in a crowded region towards the centre of the Galaxy, covering 17.4 deg^2 in $|\ell| < 5°$, $|b| < 1.5°$, result in $\sim 750\,000$ sources per colour. An investigation of the completeness and systematic effects present in these crowded images suggests that the standard DENIS magnitude limits of $\sim$18, 16 and 14 in I,J and K$_s$ were not achieved due to crowding – 80% completeness falls about 2 magnitudes brighter than these limits.

We convolve a detailed model of the systematic and random errors in the photometry with a simple model of the Galactic disk and dust distribution, to simulate expected colour-magnitude diagrams. These are in good agreement with the observed diagrams, allowing us to isolate those stars from the inner disk and bulge. After correcting for local dust-induced asymmetries, we find evidence for longitude-dependent asymmetries in the distant J and K$_s$ sources, consistent with the general predictions of some Galactic bar models. Complementary nbL-band (3.6μm) observations taken at UKIRT at $b = -0.1°$ and $\ell=\pm4.3°$ and $\ell=\pm2.3°$ are also presented. The magnitude limit ($\sim$12) and low coefficient for interstellar extinction at this wavelength ($A_{nbL}=0.047A_V$), allows us to observe bulge giants. We successfully match 95% of nbL sources with DENIS K$_s$ sources. Dereddened number counts show an excess of $\sim$15% and $\sim$5% in source counts at $\ell = -4.3°$ and $\ell = -2.3°$ compared to $\ell = +4.3°$ and $\ell = +2.3°$ respectively. This is in the same sense as predicted by many bar models, with our results favouring gas dynamical models and the recent deconvolution of surface photometry data (e.g. Binney et al. 1991; Binney et al. 1997), over earlier treatments of photometric data (e.g. Dwek et al., 1995).

N. Epchtein (ed.),
The Impact of Near-Infrared Sky Surveys on Galactic and Extragalactic Astronomy, 41-47.
© 1998 *Kluwer Academic Publishers.*

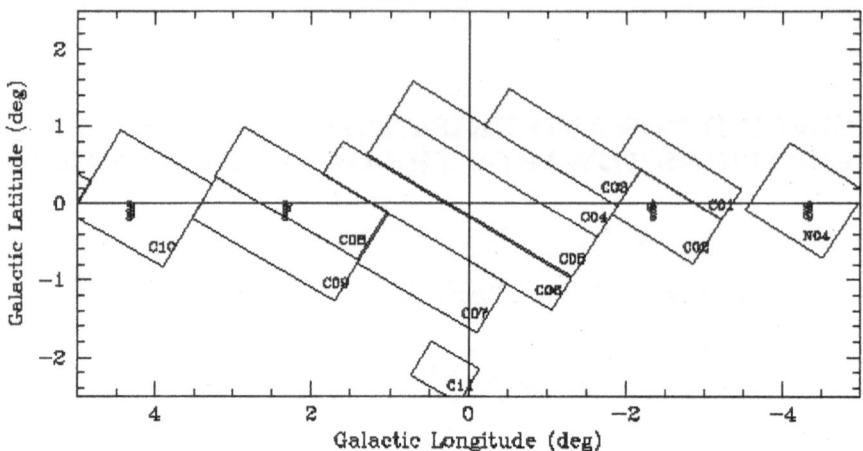

Figure 1. The observed DENIS IJK$_s$ *rasters* (large boxes). The small marked regions near $b = 0°$ and at $\ell = \pm4.3°$, $\ell = \pm2.3°$ are the locations of the supplementary nbL observations.

1. Introduction

Evidence for the existence of a kiloparsec-scale bar in the centre of our Galaxy has been accumulating. Many disparate observational techniques point to the conclusion that there is a bar with a scalelength of 2–5 kpc, with its major axis in the first quadrant – Binney et al. (1991) and Blitz & Spergel (1991) find a striking parallellogram in the observed longitude-velocity for gas motions, just as expected for motions in a rotating bar-like potential; Nakada et al. (1982) use IRAS Miras as tracers and again find the same result; OGLE event frequencies disagree with axisymmetric bulge models, but agree with bar models (Paczynski et al. 1996); COBE/DIRBE surface photometry modelling and deconvolutions (Dwek et al. 1995; Binney et al. 1997) also suggest a similar bar.

To detect the bar, however, by means of its stellar population is more difficult. We are limited by the dust which hampers the optical view of the central regions. Turning to near infra-red wavelengths, the effects of dust are much diminished. At K-band (2.2μm) the $\sim$40 magnitudes of optical extinction are reduced to only $\sim$4 magnitudes, and at 3.6μm, to only $\sim$2. The work of Hammersley et al. (1994) has used a K-band stellar survey, but was severely limited by working in only one waveband, where the effects of patchy dust clouds cannot be unambiguously disentangled from those of Galactic structure.

Multicolour surveys such as DENIS allow some attempt to be made to separate these two effects.

2. DENIS data reduction and completeness

The DENIS images used, taken between April and September 1996, cover the regions shown in figure 1. The images were prepared according to the standard pipeline reduction (Borsenberger, 1997). Each image is 768×768 pixels, which, in the case of the J and K_s images is the result of microscanning with a 256×256 detector. Source extraction and aperture photometry with an aperture of 7 arcseconds diameter were carried out using the SExtractor program (Bertin & Arnouts 1996). Standard stars observed just before and after each *raster* (see caption of figure 1) were used for photometric calibration. A thorough investigation of photometric errors and completeness was carried out. Random photometric errors were deduced by comparison of photometry in overlapping images. The addition of artificial stars to crowded DENIS images in each of the three colours, at various magnitudes, allowed a characterisation of the systematic offset of magnitude which results from flux from other sources entering the aperture. These artificial star experiments also allowed completeness levels to be deduced. Finally, two-dimensional maps of completeness levels for $J-K_s$ and $I-J$ colour-magnitude diagrams were constructed by use of a Monte-Carlo process based on the completeness levels. Faint and bright end cut-offs were imposed in I,J and K_s at 17,15,13 and 11,9,8 respectively. Absolute astrometry was carried out by a step-up process from optical sky survey images.

3. Disk Model

A very simple model to represent only the disk of the Galaxy, in the presence of a strong dust layer was constructed. The model consists of a stellar disk, with an exponential scale-length and scale-height (taken from Kent et al. 1991), together with a local luminosity function (Garwood & Jones, 1987). The dust layer is specified by its scale-length, scale-height, the strength of dust extinction, and the displacement of the sun from the plane. The model output is a list of sources generated by a streamlined Monte-Carlo method for given lines of sight, using these model parameters. The output is subsequently convolved with a parametrisation of the derived completeness limits. The dust parameters are refined by a fit to the data.

4. Bulge Models

Figure 4 shows the number count asymmetries expected according to various bar models of the Galactic central regions. For a centrally concentrated bar with its major axis in the quadrant $90° > \ell > 0°$, we expect, at relatively large longitudes ($\gtrsim 10°$) that the number counts for tracer objects should

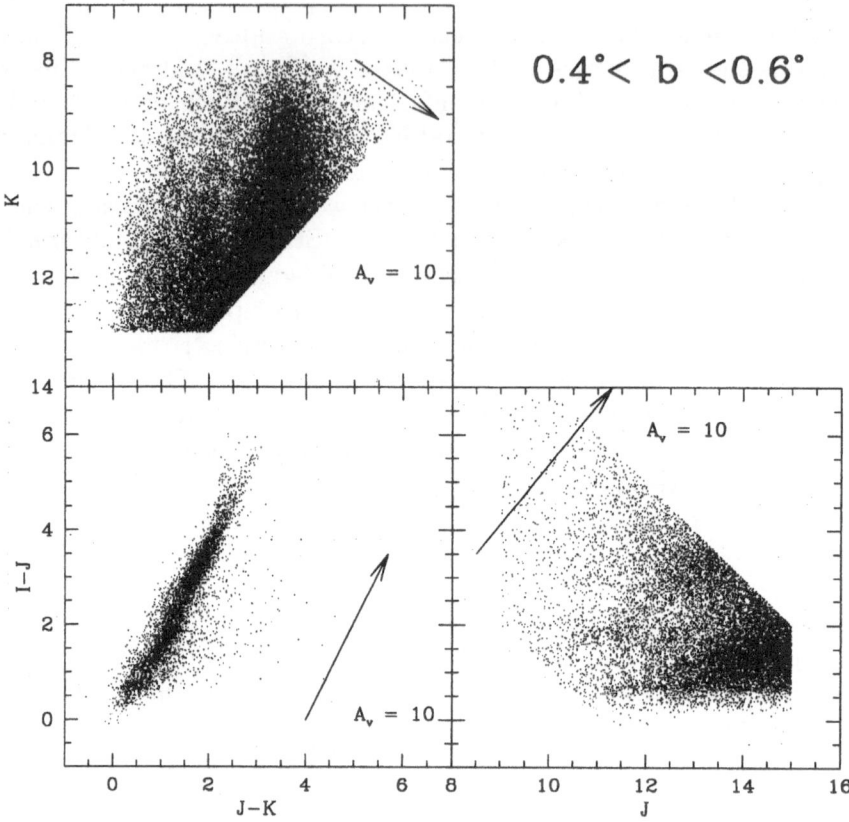

Figure 2. A typical set of colour-magnitude diagrams (for b=0.5°). Shown are (J)-(J−K), (I)-(I−J) and (I−J)-(J−K) diagrams, each constructed separately. The direction of the reddening line is shown in each diagram. Notice the rotated (I)-(I−J) axes. The most striking feature of these diagrams is the strong reddening. The limits of unreddened I−J and J−K$_s$ colours are between ∼0 and ∼1 for most main-sequence and giant stars, yet the ranges we see here are much large, and are consistent with stars seen through heavy reddening. We use the reddening as a filter to separate local and distant sources, and to investigate inner disk/bulge asymmetries.

be larger at a positive longitude +ℓ than at the corresponding negative longitude −ℓ. However, at smaller longitudes (≲ 10°), the bar geometry dictates that this contrast will be reversed with greater number counts being seen at *negative* longitudes. (see Blitz et al. 1991). We choose a few representative bar models from the literature – we use two of the best fitting models (E3 and G2) and from Dwek et al. (1995) based on a fit to DIRBE 2.2μm surface photometry; we use two dynamical models, those

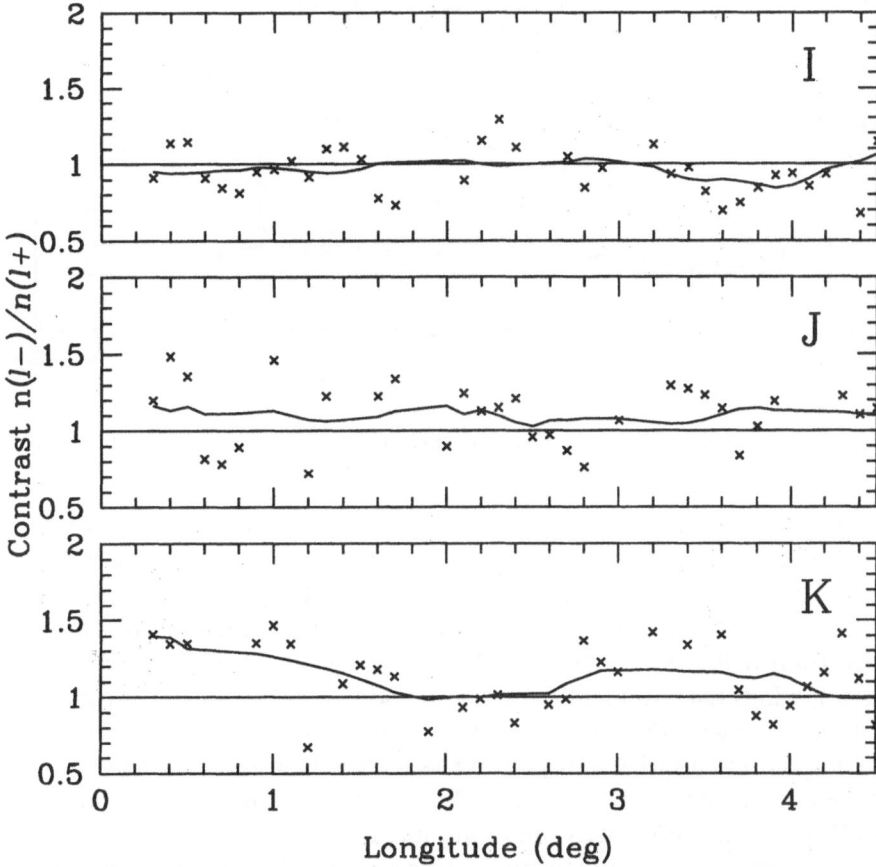

Figure 3. The expected number count asymmetries from various bar models. The points correspond to the results of the nbL observations after dereddening, with Poisson counting errors indicated.

of Binney et al. (1991) and Blitz et al. (1991); and we use a recent three-dimensional deconvolution of the DIRBE near-IR photometry by Binney et al. (1997). According to most of these models, we typically expect a peak in the number count contrast around 4–6°.

5. Counting experiment

By means of our disk model, we can delimit regions in the CM diagrams dominated by near-disk, and also by far-disk/bulge sources. At the bright magnitudes corresponding to nearby disk objects, a fit is made between equal negative and positive longitude pairs, to remove the effects of local

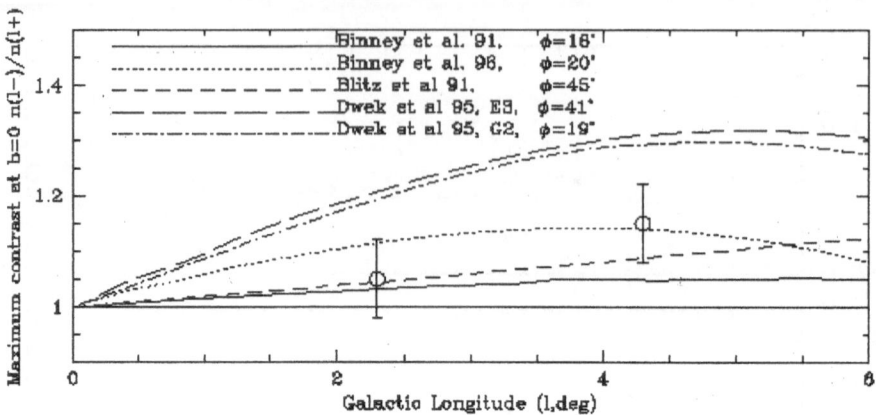

Figure 4.   The expected number count asymmetries from various bar models.

asymmetries. Any additional asymmetries remaining at fainter magnitudes will (in the case of the J band and K_s band) contain some signal of asymmetries in the inner disk or in the bulge. The I band, according to the model distances, should serve as a control, since it is expected to penetrate no more than a few kpc – the number counts seen should, after this correction for differences in local extinction, show equal values at positive and negative longitudes. This is what we see (see figure 3). The deviation of the mean ratio from unity is no more than 1.5σ. However, for the J and K_s bands, the deviation from unity is $\sim 3\sigma$. Note that the signal is not clean. The method we use, although removing local asymmetries in the disk, does not remove asymmetries which may be caused by dust nearer the centre.

6. nbL-band data

The noisy results from IJK$_s$ data led to the use of a wavelength with an even lower coefficient of extinction. It was clear, according to the model distances, that while DENIS K–band observations were likely to penetrate, DENIS J-band observations might not reliably penetrate the most severe extinction towards the central regions of the galaxy.

Thus complementary nbL (3.6μm) observations were made covering a total of 277 arcmin^2 equally distributed between fields at $\ell = \pm 4.3°$ and $\ell = \pm 2.3°$ were taken at UKIRT in July 1996. (The field positions are shown in figure 1) The observation mode required the co-adding of multiple very short exposures, since the sky background at 3.6μm is very bright and easily saturates the detectors. Data reduction was carried out in the standard way, with a flat field constructed by the median filtering of several survey images. The small pixel size (0.286 arcseconds) meant that severe confusion, as in the DENIS images, was not a problem. The astrometry was carried out by matching images to the DENIS K_s catalogue (see above). The resulting 95% match of K_s and nbL sources suggests that the penetration by neither K_s and L band data is severely limited by dust. We deredden the stars by assuming the majority of the stars to be K_s and M giants displaced at 8 kpc. A calculation based on a simple galactic model (Binney et al. 1997) suggests that the inner Galaxy (within 3 kpc from the centre) has a 10–12% disk contamination for $|b| < 0.1°$. The difference in distance to the two sides of the putative bar for lines of sight at $\ell = \pm 4°$ is less than a few tenths of a magnitudes, and we use the dereddened nbL-band stars as tracers. The resulting number count contrast is shown in figure 4. We see that the gas dynamical (Blitz et al 1991; Binney et al. 1991) and the more recent photometric model (Binney et al. 1997) is preferred over the earlier photometric work by Dwek et al. (1995).

References

Bertin E., Arnouts S., 1996, A&AS, 117, 3

Binney J., Gerhard O.E., Stark A.A., Bally J., Uchida K.I., 1991, MNRAS, 252, 210

Binney J., Gerhard O., Spergel D., 1997, MNRAS, 288, 365

Blitz L., Spergel D.N., 1991, ApJ, 379, 631

Borsenberger J., 1997, in *The Impact of Large Scale Near-IR surveys*, eds. F.Garzón et al., (Kluwer), p181

Dwek E. et al., 1995, ApJ, 445, 716

Freudenreich H.T., 1996, ApJ, 468, 663

Garwood R., Jones T.J., 1987, PASP, 99, 453

Hammersley P.L., Garzón F.,Mahoney T., Calbet X., 1994, MNRAS, 269, 753

Kent S.M., Dame T.M., Fazio G., 1991, ApJ, 378, 131

Nakada Y., Deguchi S., Hashimoto O., Izumiura H., Onaka T., Sekiguchi K., Yamamura I., 1991, Nature, 353, 140

Paczynski B. et al., 1996, in Blitz L., Teuben P., eds., IAU 169, *Unsolved Problems of the Milky Way*, Kluwer, Dordrecht, p93

MASS-LOSING AGB STARS AND YOUNG STARS IN THE ISOGAL SURVEY

A. OMONT AND B. CAILLAUD,
Institut d'Astrophysique de Paris-CNRS, France

P. FOUQUE,
Observatoire de Paris, DESPA, France and European Southern Observatory, Chile

G. GILMORE,
Institute of Astronomy, Cambridge, U. K.

D. OJHA,
TIFR, Bombay, India

M. PERAULT, AND P. SEGUIN,
Laboratoire de Radioastronomie Millimétrique, ENS and CNRS, Paris, France

G. SIMON,
Observatoire de Paris, DASGAL, France

R.F.G. WYSE,
John Hopkins University, Baltimore, USA

AND

THE ISOGAL-DENIS TEAM

ABSTRACT

ISOGAL is the first near+mid-infrared, high-resolution imaging survey of the inner disk and bulge of the Milky Way Galaxy (Pérault et al. 1996). The ISO satellite, and especially ISOCAM, allows a quantum jump in observational analyses of Galactic structure and stellar populations, star formation and infrared properties of the interstellar medium, especially in the most obscured regions of the inner Galactic disk. The combination of ISOCAM data with DENIS (Epchtein et al. 1997) IJK data allows characterisation of the stellar populations.

N. Epchtein (ed.),
The Impact of Near-Infrared Sky Surveys on Galactic and Extragalactic Astronomy, 49-56.
© 1998 *Kluwer Academic Publishers.*

We illustrate the scientific capabilities of ISOGAL with an analysis of the stellar sources in a field in the inner Galactic bulge/disk ($\ell, b = 0, +1$). ISOGAL+DENIS has sufficient sensitivity resolution to be almost complete to the RGB tip in the inner Galactic bulge at all of $2\mu m$, $7\mu m$ and $15\mu m$, allowing analysis of bulge mass-losing AGB stars. In this field the relative number of mass-losing stars with respect to non-mass-losing AGB stars of similar luminosity is larger than in the solar neighbourhood. The scientific ramifications of these observations are uncertain – is this a metallicity effect?

The study of stellar formation at Galactic scales is another major goal of ISOGAL. The analysis of the first results confirm the expectations derived from the IRAS results on YSOs in nearby giant molecular clouds: ISOGAL is able to detect solar mass YSOs (of "type I and II") up to a few kpc, and those of intermediate mass up to the Galactic Center. The combination with the near infrared data of the DENIS survey allows to discriminate YSOs from other ISOGAL sources, mainly red giants. From the detection of more than 10 000 YSOs, one can expect a detailed and global view of star formation in the different environments of the most obscured regions of the inner Galaxy.

1. WHAT IS ISOGAL?

ISOGAL is the first near+mid-infrared, high-resolution imaging survey of the inner disk and bulge of the Milky Way Galaxy, combining ISOCAM $7\mu m$ (LW2) and $15\mu m$ (LW3) images with DENIS IJK_s images. These data allow the determination of the distribution of interstellar extinction on small angular scales, and hence deduction of intrinsic stellar colours and spatial distributions, thereby facilitating a broad range of analyses of Galactic structure. Additionally, from the reddening-corrected photometric data, one may search for and identify any remarkable individual stars. The ISOCAM sensitivity allows us to detect dusty low-mass young stars up to several kpc from the Sun, and regions of current or recent star formation, and individual bright M giants, to beyond the Galactic Center. The central parts of the Milky Way are of considerable intrinsic interest; the inner disk and bulge contain most of the stars and interstellar medium, most active star-formation regions, the greatest number of products of the late stages of stellar evolution, most rare and short-lived stages of stellar evolution (which can dominate the infrared luminosity of galaxies), and exotic objects. The most important manifestations of spiral arms and of large-scale dynamical asymmetries may be studied in unique detail, and with unprecedented reliability.

With these goals in mind, we have selected fields distributed along the inner Galactic disk, to optimise the range of studies outlined above – the inner Galactic bulge is densely sampled, with less dense sampling of the disk out to longitude $|\ell| = 60°$. We discuss here the results from two ISOGAL fields of the inner Galaxy, with an emphasis on the RGB and AGB stars and young stars. The results illustrate the capabilities of the ISOGAL data and provide an indication of the scientific returns from this survey.

2. THE DATA

The fields analysed here are the one at $\ell = -45°$ described in Pérault et al. (1996), and a small field at $(\ell, b) = (0.0, 1.0)$ with area 0.038 sq deg.

The DENIS source density is approximately 50000 per sq deg, being confusion limited (~ 30 pix/source) by the DENIS 3″ pixels. The completeness limits are K$\sim$10.5, J$\sim$12.5 and I$\sim$13.5. The K/J-K colour-magnitude diagramme (Fig.1) shows a well-defined bulge red giant sequence, shifted by fairly uniform extinction of $A_V = 6 \pm 1$ mag. with respect to the reference K_o vs $(J - K)_o$ of Bertelli et al. 1994) with Z=0.02.

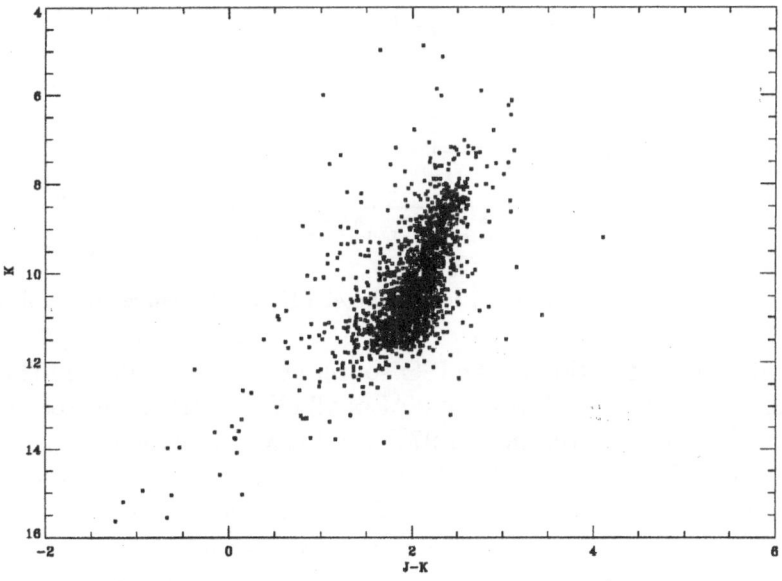

Figure 1. K versus $J - K$ magnitude color diagram of the DENIS sources

ISOGAL observed this field with the broad filters LW3 (12-18 microns) and LW2 (6-8.5 microns) with 6″ pixels in a (4×7) raster. Since this field is one of the smallest observed by ISOGAL, edge effects are relatively large, with only $\sim 60\%$ of the field observed twice. The images have been processed by the ISOCAM team IA software, with some enhancements. Source extraction and photometry is based on a maximum-likelihood analysis optimised for crowded fields and under-sampled data.

Source counts are displayed in Figures 2 and 3. With source densities typically ~10000 per sq. deg. (~40 pixels per source), point-source identification is again confusion limited. Quantification of ISOCAM data completeness is underway.

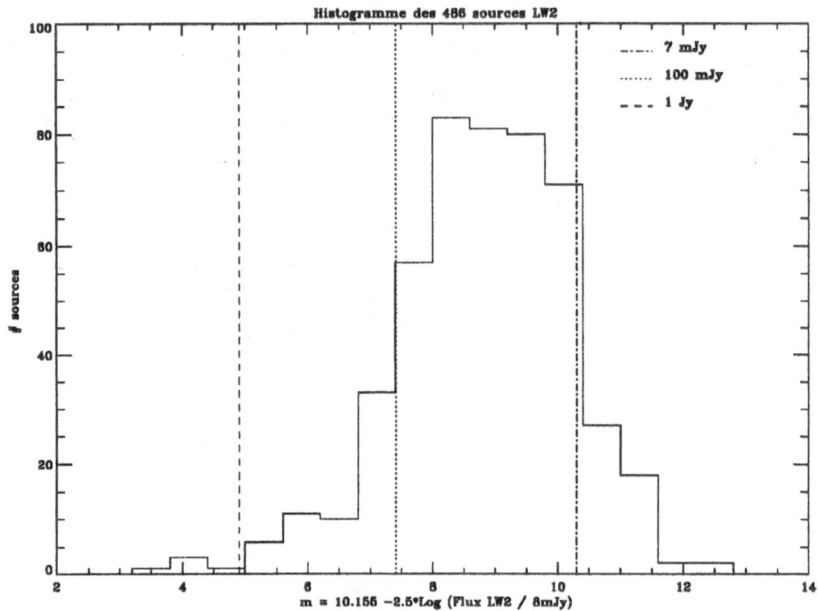

Figure 2. Magnitude distribution of the 486 LW2-ISO sources (central field)

A substantial fraction of the ISO sources can be identified in the DENIS J and K data: 85% of all LW3 and 90% of all LW2 sources are cross-matched, with the fractions increasing to 97% for LW3 sources with $S_{15} > 10$ mJy and 98% for LW2 sources with $S_7 > 15$ mJy. Note also that ~80% of LW3 sources with $S_{15} > 5$ mJy have an LW2 counterpart.

We now restrict discussion to point sources above the completeness limits, considering the 216 sources detected in both the LW3 and LW2 bands with $S_{15} > 5$ mJy and $S_7 > 7$ mJy.

3. THE MANY MASS-LOSING AGB STARS IN THE INNER BULGE/DISK

The striking feature of the colour-colour and magnitude-colour diagrams derived in the $\ell=0$ b=1 field from the ISOGAL and DENIS data of the 216

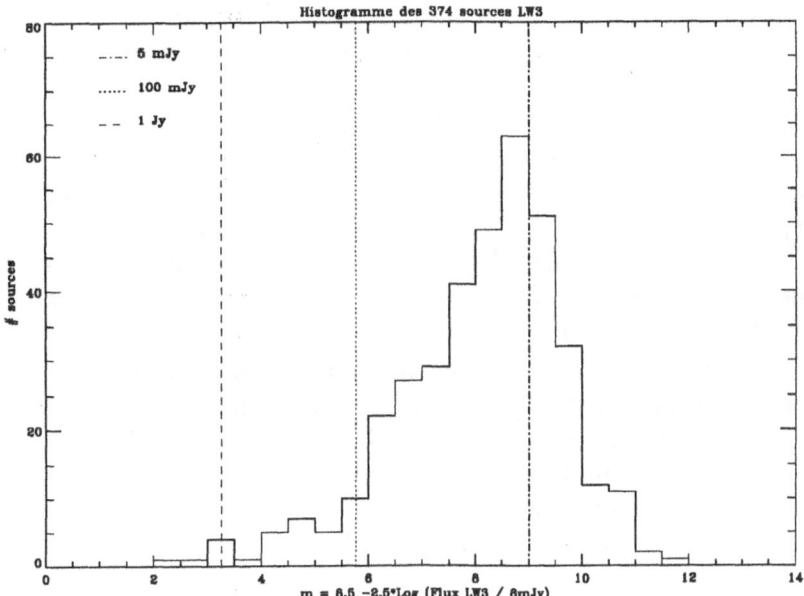

Figure 3.   Magnitude distribution of the 374 LW3-ISO sources (central field)

bright sources is the large range of K-[15] and K-[7] colours. Interstellar extinction in this line of sight is $A_K \sim 0.6$, $A_7 \sim 0.2$ and $A_{15} \sim 0.1$. The photospheric colours of normal late-type red giants are $(K-[15])_o < 0.3$ and [7]-[15]< 0.3. More than 40% of the sources under discussion here have observed K-[15] colours redder than K-[15]$\sim$1, and very red [7]-[15] and K-[7]colours. Thus these data suggest these are mass-losing stars.

Further, most of these bright ISO sources have apparent K magnitudes ($K_o + 0.6$) which are above the RGB tip ($K_{o,tip} \sim 8.3$ in the bulge, see e.g. Tiede et al. 1995). Almost all the very red stars are above the RGB tip, and are thus AGB stars. What fraction of all AGB stars are this red? Our LW3 completeness limit is such that we would detect only $\sim$ 50% of non-mass-losing AGB stars just above the RGB tip. However, DENIS is complete in K to significantly below the RGB tip. Thus, the number of non-red excess AGB stars this luminous in K is derivable from the $\sim$ 300 DENIS sources with K< 9 (foreground disk contamination is negligible in this field).

We conclude that at least 25% of the AGB stars above the RGB tip are losing mass, from their very red ISO colours. This high proportion of mass-losing red giants above the RGB tip is larger than the value in the solar neighbourhood (see e.g. Wainscoat et al. 1992 and Glass et al. 1995)).

Our conclusion of a large mass-losing fraction is robust, since the large spread of ISOGAL colours is certainly real. However the exact source counts are sensitive to our completeness corrections, which remain under development.

AGB models (Justtanont and Tielens 1992, M. Groenewegen in preparation) suggest that our observed colours correspond to mass-loss rates of order a few $10^{-7} M_\odot yr^{-1}$. The $\sim 10\%$ of the sources with the largest $15\mu m$ excesses probably have larger mass-loss rates, of order a few $10^{-6} M_\odot yr^{-1}$. These most extreme sources also have detectable J-K excess, with a correlation between J-K colour and K-[15] colour, further establishing the physical validity of our identification of them as AGB stars with large mass-loss.

We have begun to analyse the red giant population in other ISOGAL fields. For instance in the symmetric field (ℓ=0 b=-1), the content of red giants seems quite similar, but the extinction is much larger and variable which makes the estimate of intrinsic colours more difficult. Farther in the bulge (b=-2.17), their density is about twice smaller while their infrared colours look relatively similar. In the central disk (b$\sim$0°), the extinction is even larger (up to $A_v \sim$35).

A first analysis of a field at ℓ=-45° has been given in Pérault et al. (1996). About half (180) of the 395 $15\mu m$ sources have a clear identification with a bright K source (K < 11) with a matching radius 6". Most of them have a J counterpart. The colour-colour diagrams (J-K)/(K-[15]) (Pérault et al. (1996) and (J-K)/(K-[7]) show that most of these bright K sources are luminous red giants with a large interstellar extinction (up to $A_v \sim$30). Many of them should be mass-losing AGB stars.

4. DUSTY YOUNG STARS

In this field at ℓ=-45° in addition to such relatively well defined sequence of red giants, there are also a number of 15 μm sources associated with relatively faint K sources and large values of K-[15], or without any K association (Pérault et al. 1996).

With a 6" matching radius, 52 ISOGAL sources are associated with a weak DENIS K source (12.5>K>11), among them 27 have also a J counterpart. Most of these sources have large values of K-[15] and relatively small values of J-K, i.e. typical colours of T Tauri stars. However, more than half of them should be pure chance associations, and one has to wait for a better ISOGAL astrometry to confirm the nature of the others. The number of 15 μm sources without K association is $\sim$160. Thus almost half the total number of 15 μm sources have no K association. The only plausible candidates for such a large number of cold sources are dusty young stars. The expected numbers of detectable galaxies and asteroids in our 0.144 deg^2 field are less than 1 and 5 respectively.

As discussed by Pérault et al. (1996), the observed density of 15 μm sources without K counterpart ($\sim$1300/deg^2) is just in the range of the expected number of young sources in a giant molecular cloud (GMC) a few kpc away, inferred from the IRAS results in nearby clouds such as L1641 in Orion (Strom et al. 1989). From this detailed study of Strom et al., it appears that "Type II" YSOs should have typically K-[15]>4, while "Type I" YSOs should have K-[15]> 8. Most similar objects detected at 15μm should not be detectable in the K band at 3-5 kpc with our sensitivity. But many class II sources could be detectable with a slightly better sensitivity, as possibly observed in another ISOGAL field (Testi et al. in preparation) with TIRGO data ($K_{LIM} \sim$ 14.5).

In conclusion, the ISOGAL survey presents a major interest for the study of stellar formation at Galactic scales. The analysis of the first results confirm the expectations derived from the IRAS results on YSOs in nearby giant molecular clouds: ISOGAL is able to detect solar mass YSOs (of "type I and II") up to a few kpc, and those of intermediate mass up to the Galactic Center. The combination with the near infrared data of the DENIS survey allows to discriminate YSOs from other ISOGAL sources, mainly red giants. From the detection of more than 10 000 YSOs, one can expect a detailed and global view of star formation in the different environments of the most obscured regions of the inner Galaxy: disk, disk-bulge transition region, bar, molecular ring, spiral arms, inter-arm foreground regions, etc.

However, the reality of all the weak sources detected by ISOGAL without K association still needs to be confirmed, and their nature should be determined by follow-up observations such as deeper K images (used with an improved ISOCAM astrometry) and visible and infrared spectrometry. It should also be stressed that in addition to such individual YSOs, ISOGAL will also bring various other information related to star formation in the inner Galaxy: (young) open clusters and other groups of young stars, supergiants, young bright red giants, rich diffuse IR emission by PAHs and dust in regions of stellar formation, dark cloud condensations of various sizes traced by high IR extinction, etc.

References

Bertelli G. et al. 1994 A&AS 106, 275

Epchtein N. et al. Messenger 87, 27

Glass I.S., Whitelock P.A., Catchpole R.M. & Feast M.W. 1995 MNRAS 273, 383

Justtanont K. & Tielens A.G.G.M. 1992, ApJ 389, 400

Pérault M., Omont A., Simon G., Séguin P., Ojha D., Blommaert J., Felli M., Gilmore G., Guglielmo F., Habing H., Price S., Robin A., de Batz B., Cesarsky C., Elbaz D., Epchtein N., Fouqué P., Guest S., Levine D., Pollock A., Prusti T., Siebenmorgen R., Testi L., Tiphène D., 1996, AA 315, L165

Strom K.M., Newton G., Strom S.E., Seaman R.L., Carrasco L., Cruz-Gonzalez I., Serrano A. 1989, ApJS 71, 183

Tiede G.P., Frogel J.A. & Terndrup D.M. 1995, AJ 110, 2788
Wainscoat R.J. et al. 1992, ApJS 83,111

GALACTIC BULGE FROM TMGS STAR COUNTS

M. LÓPEZ–CORREDOIRA, F. GARZÓN, P. L. HAMMERSLEY,
T. J. MAHONEY AND X. CALBET
Instituto de Astrofísica de Canarias
E-38200 La Laguna, Tenerife, Spain

Abstract. The bulge of the Galaxy is analysed by inverting K-band star counts from the Two-Micron Galactic Survey in a number of off-plane regions. Assuming a non-variable luminosity function within the bulge, we derive the top end of the K-band luminosity function, which shows a sharp decrease brighter than $M_K = -8.0$ when compared with the disc population, and the stellar density function, whose morphology is fitted to triaxial ellipsoids with the major axis in the plane at an angle with line of sight to the Galactic centre of 12° in the first quadrant. The axial ratios are 1 : 0.54 : 0.33 and the distance of the Sun from the centre of the triaxial ellipsoid is 7860 pc.

We use some off-plane regions in the Galactic bulge ($10° > |b| > 2°$, $|l| < 15°$) taken from TMGS (Two-Micron Galactic Survey; Garzón et al. 1993) up to nineth K magnitude. The total sky coverage is some 75 deg^2 of sky. In the areas considered, the contribution to the star counts will be primarily from the disc and bulge and the extinction is low (between 0.05 to 0.5 mag at K) and not too patchy (Garzón et al. 1993).

In order to isolate the bulge component a model disc was subtracted from the total counts: Wainscoat et al. (1992; hereafter WCVWS), which has been used because it provides a good fit to the TMGS counts in the region where the disc dominates (Cohen 1994). So $N_{K,\text{bulge}}(m_K) = N_K(m_K) - N_{K,\text{disc}}(m_K)$ and (in rad^{-2})

$$N_{K,\text{bulge}}(m_K) = \int_0^\infty \Phi_{K,\text{bulge}}(m_K + 5 - 5\log_{10} r - a_K(r)) D_{\text{bulge}}(r) r^2 dr, \quad (1)$$

where $\Phi_{K,\text{bulge}}(M) = \int_{-\infty}^M \phi_{K,\text{bulge}}(M)dM$, ϕ is the normalised luminosity function ($\int_{-\infty}^\infty \phi(M)dM = 1$), D is the density, and a_K is the extinction in the line of sight according to WCVWS.

N. Epchtein (ed.),
The Impact of Near-Infrared Sky Surveys on Galactic and Extragalactic Astronomy, 57-61.
© 1998 *Kluwer Academic Publishers.*

With the change of variables $\rho_K = 10^{0.2a_K(r)}r$ and $\Delta_K = D(r)\frac{r^2 dr}{\rho_K^2 d\rho_K}$ we transform the equation (1) of counts in the bulge into

$$N_K(m_K) = \int_0^\infty \Phi_K(m_K + 5 - 5\log_{10}\rho_K)\Delta_K(\rho_K)\rho_K^2 d\rho_K. \qquad (2)$$

The density is obtained by inverting this equation: Δ is the unknown function and Φ is the kernel of a Fredholm integral equation of the first kind.

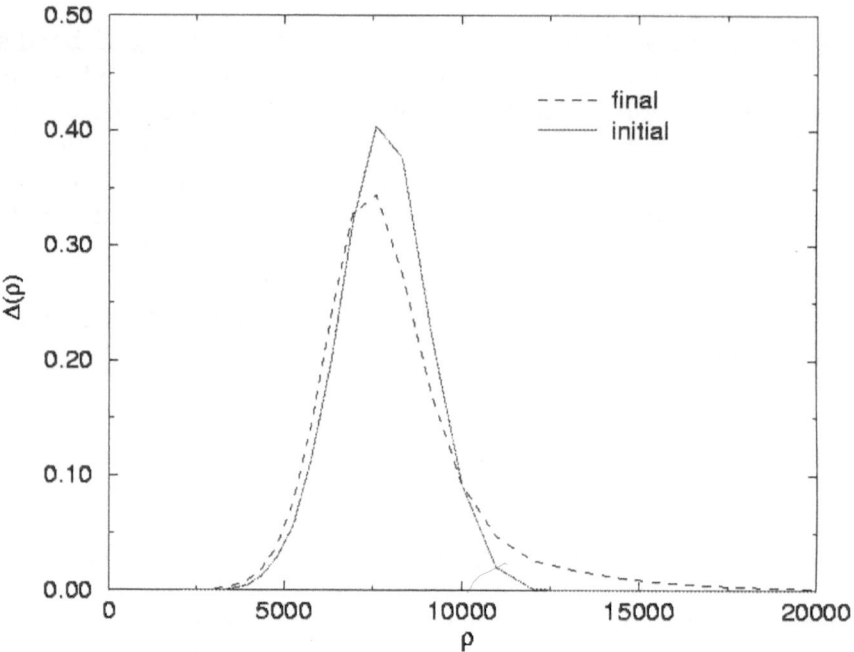

Figure 1. Recovery of the theoretical luminosity function through the inversion proccess.

When the luminosity function Φ, is the unknown instead of Δ, then we can make a new change of variable $M_K = m_K + 5 - 5\log_{10}\rho_K$ and we obtain a new first kind of Fredholm equation:

$$N_K(m_K) = 200(\ln 10)10^{\frac{3m_K}{5}} \int_{-\infty}^\infty \Delta_K(10^{\frac{5+m_K-M_K}{5}})10^{\frac{-3M_K}{5}}\Phi_K(M_K)dM_K,$$
$$\qquad (3)$$

where Φ is now the unknown function and Δ_K is the kernel.

Both integral equations are inverted using Lucy's statistical method (Lucy 1974). This method is fairly insensitive to the high-frequency fluctuations and in our tests with known functions, which are similar to that of the bulge, gave good results (note: this method would not be applicable to the disc as a whole). Fig. 1 shows the outcome of a numerical simulation where we introduce an initial Δ function, obtain the counts through (2) adding a Poissonian equivalent noise over a square degree of sky and invert by Lucy's algorithm: the final Δ is similar to the initial Δ in spite of noise.

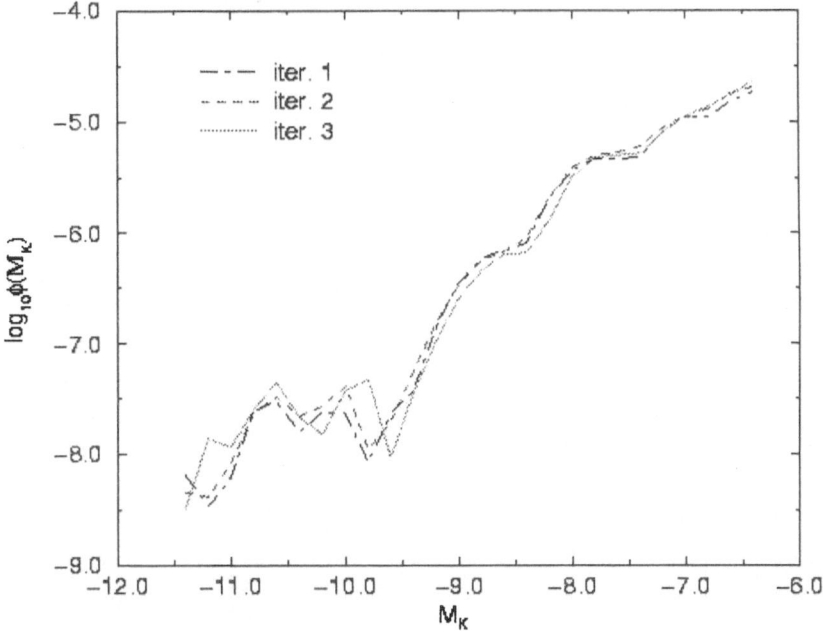

Figure 2.   Luminosity function in the first three iterations.

We solved first for the luminosity function for each region using the WCVWS bulge density. Then, assuming that the luminosity function is independent of the position (while our assumption may not be not strictly true, it is still a reasonable approximation between about 250 pc and 1200 pc from the galactic plane) we average it and use for inverting (2) to derive a new density distribution. The new density was then used to improve the luminosity function, etc. The whole process was iterated three times which was enough for the results to stabilize as can be seen in the Fig. 2: we see how the result of the third iteration is very close to the first; i.e. stabilization is reached in the first iterations.

The obtained luminosity function is shown in the Fig. 3 and it is accurate enough (the number of bulge stars is large enough compared with possible sources of contamination) down to $M_K \sim -9.5$. The coincidence of our luminosity function with that of WCVWS for the faintest stars is due to the fact that we used their luminosity function to initiating the iteration process. Practically speaking, this overlapping corresponds to an effective normalization to the WCVWS luminosity function in the range $M_K > -7$.

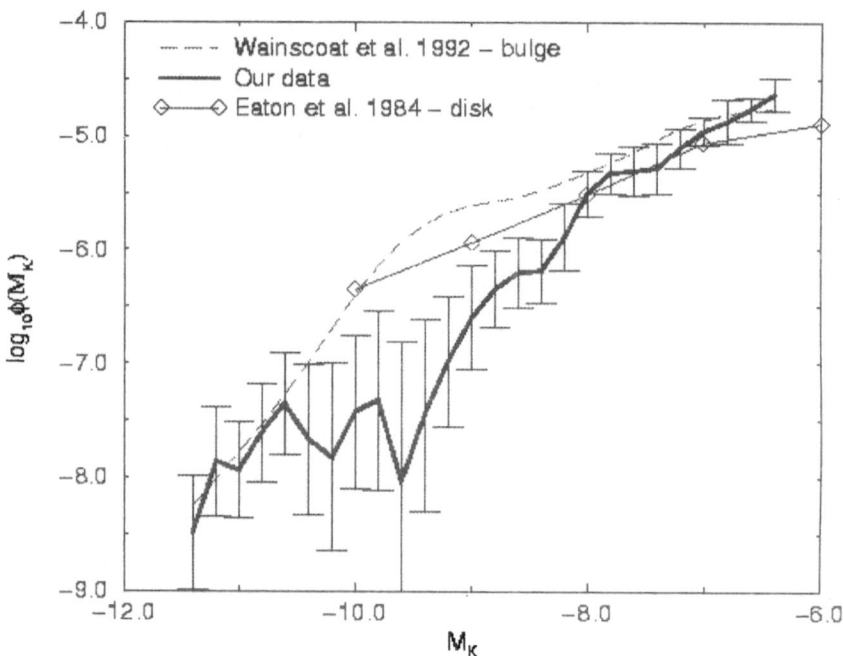

Figure 3. Luminosity function of the bulge stars in the K band: the thick solid line is the third and final iteration. Comparisons with the same luminosity function in the first iteration, WCVWS in the bulge and Eaton et al. (1984) in the disc are also provided.

Fig. 3 shows that for $-10 < M_K < -7$ the bulge luminosity function is significantly lower than that of the disc (Eaton et al. 1984). The discrepancy with WCVWS could arise from their not having taken into account that the brightest stars in the bulge are up to 2 magnitudes fainter than the disc giants (Frogel & Whitford 1987). This would shift the luminosity function to the right.

The morphology of the bulge can be examined by fitting the isodensity surfaces to $D(\vec{r}) = D(r, l, b)$. We fitted three-dimentional ellipsoids with two axes in the Galactic plane, x and y (we have ignored a possible tilt out of the plane), to 20 isodensity surfaces (from 0.1 to 2.0 star pc^{-3}, in steps of 0.1) with four free parameters: R_0, the Sun-Galactic centre distance; K_z and K_y, the axis ratios with respect the major axis x; and α the angle between the major axis of the triaxial bulge and the line of sight to the Galactic centre.

The four averaged parameters have been fitted for the 20 ellipsoids and the results are: $R_0 = 7860 \pm 90$ pc, $K_z = 3.0 \pm 0.9$, $K_y = 1.87 \pm 0.18$ and $\alpha = 12 \pm 6$ deg. We can also express the axial ratios as $1 : 0.54 : 0.33$. The errors are calculated from the average of the ellipsoids, and so do not include possible systematic errors. Hence, the true errors are larger that stated but tests suggest that they do not alter the general findings of this paper. These numbers indicate that the bulge is triaxial with the major axis close to the line of sight towards the Galactic centre. The error in K_z is quite large and is due to a non-constant axial ratio of the ellipsoids. There is a trend towards increasing K_z with proximity to the centre, i.e. the outer bulge is more circular than the inner bulge.

References

Cohen M., 1994, Ap&SS, 217(1), 181
Eaton N., Adams D. J, Gilels A. B., 1984, MNRAS, 208, 241
Frogel J. A., Whitford A. E., 1987, ApJ, 320, 199
Garzón F., Hammersley P. L., Mahoney T., Calbet X., Selby M. J., Hepburn I. D., 1993, MNRAS, 264, 773
Lucy L. B., 1974, AJ, 79(6), 745
Wainscoat R. J., Cohen M., Volk K., Walker H. J., Schwartz D. E., 1992, ApJSS, 83, 111 (WCVWS)

IR STAR COUNTS IN THE INNER DISC

P.L. HAMMERSLEY, F. GARZÓN, T. MAHONEY AND
M. LÓPEZ-CORREDOIRA

Instituto de Astrofísica de Canarias E-38200 La Laguna

Abstract. New small-scale JH star counts are presented for 4 regions on the plane between $l=31°$ and $l=15°$. The colour-magnitude diagrams clearly show that there is a major giant branch at $l=27°$ which is not present at $l=31°$. The distance to this feature is about 7.5 kpc with a total extinction of $A_V=7$ mag. This giant branch is also seen at $l=21°$ and $l=15°$ and is consistent with there being a major bar at an angle of about 75° to the line of sight. At $l=31°$ only features attributable to the disc can be seen and show that the extinction within the molecular ring is significantly higher than along the rest of the line of sight.

1. Introduction

Currently there is no satisfactory explanation for the distribution of stars near the plane for $l=33°$ to 15° and $l=-15°$ to $-30°$. Between $l=27°$ and 15° the Two Micron Galactic Survey (TMGS, Garzon et al. 1992) is finding a high concentration of very luminous sources within about 0.5° of the plane, which is not seen in any other area surveyed by the TMGS (Hammersley et al. 1994). This is clearly a young population related with the inner Galaxy. The more usual suggestion is that it could be a ring. However, the DIRBE 2.2 and 3.5 μm data show that a uniform circular ring is not feasible, and that if it were a ring it must be highly elliptical. Furthermore, the TMGS data would imply that this elliptical ring only has a significant density of stars near the tangential line of sights, with very few stars towards $l=7°$ or $l=-1°$. In Hammersley et al. (1994) the problems with the ring were discussed and it was suggested that this feature could be related to the bar, however more information was required.

63

N. Epchtein (ed.),
The Impact of Near-Infrared Sky Surveys on Galactic and Extragalactic Astronomy, 63-68.
© 1998 *Kluwer Academic Publishers.*

TABLE 1. The lines of sight for areas covered. All lines of sight cross the disc and the Sagittarius and Scutum spiral arms

Position	Crosses		Comment
	stellar bar	3 kpc dust ring	
$l=31°$			Nearly tangential to the Scutum arm
$l=27°$	X		End of bar
$l=21°$	X	X	
$l=15°$	X	X	

2. The Data

Maps at J and H were made of number of small areas in the Galactic plane ($b = 0°$) at strategic positions determined from the TMGS and the DIRBE maps (Table 1). CAIN, the facility IR camera, on the 1.5 m TCS (Tenerife) was used to obtain the images. The 5 σ limiting magnitudes were about $H=16$ mag and $J=16$ mag, and the pixel scale, of 0.4" per pixel, was sufficient for confusion not to be a significant problem. In Figure 1 is presented the $J-H$ vs. H colour-magnitude diagrams for the regions at $l=31°, 27°, 21°$ and $15°$. The area covered at each region is given in the plots.

In order to help interpret these colour-magnitude diagrams Figure 2 shows the $l=31°$ region with the position of giants and dwarfs for various distances and total extinctions along the line of sight marked.

3. Results

The main difference between the $l=31°$ and $27°$ regions is the presence of a major giant branch in the latter region.

3.1. THE DISC

The $l=31°$ region is dominated by disc sources. The dwarf stars are less luminous than the giants and even with the magnitude limit of around $H=+16$ mag, the maximum distance for a G dwarf is around 2 kpc. Therefore the amount of reddening will be limited and hence the dwarfs form the clump with $J-H$ between 0 and 1.

The disc giants, which have their peak space density near K0III to K3III, form the diagonal stripe which can be seen in the $l=31°$ region between $(J-H=1,H=12)$ and $(J-H=2,H=14)$. This feature is visible in all the areas, although the actual gradient depends on the extinction. However this stripe does not form a smooth shallow curve as is predicted by the model;

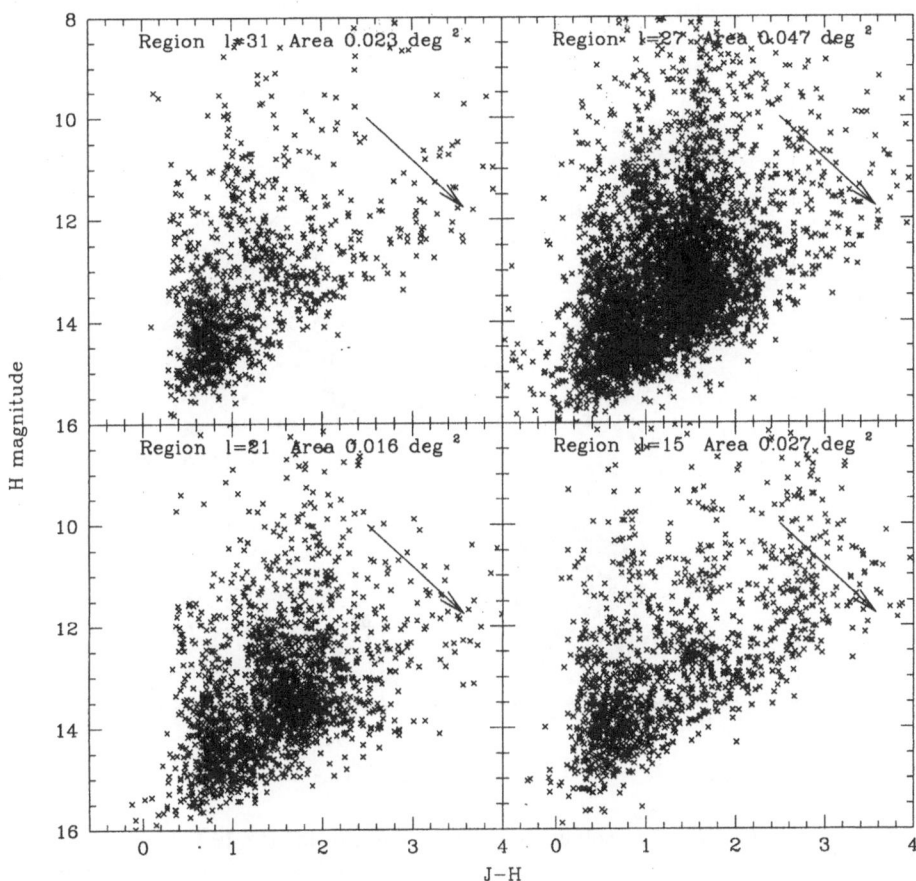

Figure 1. The $J-H$ vs. H colour-magnitude diagram for the regions at $l=31°, 27°, 21°$ and $15°$. The arrow indicates the reddening vector for 10 magnitudes of visual extinction.

rather, there is a significant dog-leg at $(J-H=1.2, H=12.5)$. The only reasonable explanation for this is that the level of distributed extinction along the line of sight jumps suddenly by a factor three. If it is assumed that the stars causing the stripe are K2III then the dog-leg is at a Galactocentric distance of about 5 kpc (assuming $R_\mathrm{o}=8$ kpc) and extends inwards to at least 4 kpc (the data is not sufficiently deep to determine what happens inside this distance). The location of this increased extinction is coincident with the molecular ring, which shows up prominently in radio maps (see e.g. Clemens, Sanders & Scoville 1988). The increase in extinction is consistent with the recent result from the analysis of DIRBE data (Sodroski et al.

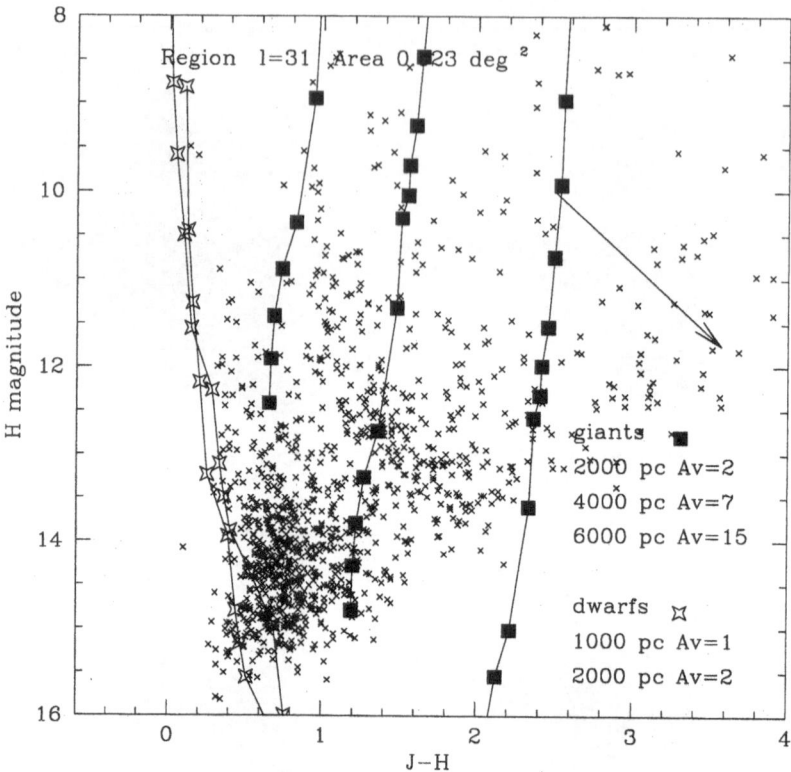

Figure 2. The $J - H$ vs. H colour-magnitude diagram for the $l=31°$ regions with the position of the dwarfs and giants at various distances and extinctions. The faintest giant marked in each case is G5III. The arrow indicates the reddening vector for 10 mag of visual extinction.

1997).

3.2. THE GIANT BRANCH

The giant branch at $l=27°$ and $l=21°$ runs from about ($J-H=1.6$, $H=14$) to ($J-H=1.8, H=8$). There are a number of important features

- From the $J-H$ colour, the total extinction to the feature is about $A_V=7$ mag.
- Assuming that the disc stripe is formed by K2III then the distance to the feature is about 7.5 kpc.
- The small spread in $J-H$ for sources within the giant branch indicates that there is little extinction within the feature, and that the feature does not extend significantly along the line of sight. The spread that

does exist can be attributed principally to the errors in the measured magnitudes.

— Over the H magnitude range covered by the plot the giant branch contributes some 50% sourced detected at $l=27°$.

— the giant branch is present at $l=21°$ and $15°$ and so extends significantly along the plane.

— The H star counts at $l=27°$ and $l=21°$ fainter than $H = +10$ are identical within the errors, whereas there are far more young stars at $l=27°$ than $l=21°$ (Hammersley et al. 1994).

4. Discussion

The giant branch indicates the presence of a major old feature in the inner Galaxy. However its location and extent are consistent with the very young feature discussed by Hammersley et al. (1994) and strongly suggest that the young and old sources stem from the same feature in the inner galaxy.

There is no evidence for either the giant branch or the excess of very young stars at $l=31°$ which is almost tangential to the Scutum spiral arm. Hence this feature is definitely not related to a spiral arm or young stars trailing off the arms, as $l=31°$ would be the position where it should be strongest. Similarly, the lack of the giant branch at $l=31°$ rules out the thick ring proposed by Kent, Dame and Fazio (1991), which would require the ring to extend to about $l=35°$.

Both the bar proposed in Hammersley et al. (1994) and an elliptical thin ring with the tangential points at $l=27°$ and $l=-22°$ could explain the presence of the giant branch and young stars at the same location, However, a bar and a ring should produce a difference distribution for the young and old star.

— A bar would be expected to have the young stars concentrated towards the end of the bar with far fewer in the middle of the bar (i.e. more at $l=27°$ than at $l=21°$), whereas the older stars would be spread more evenly down the bar.

— In principle a ring would have a similar distribution for both the old and young stars; i.e. a major peak for both young and old sources at the tangential point but falling rapidly inwards.

The evidence from this data set supports a bar far more than a ring. The bar would be expected to give the distribution of young and old sources that is seen. Only a very *ad hoc* patchy arrangement of young stars in a ring could give observed distribution.

It is now widely acknowledged that the Galaxy is barred. However, the bar that is normally referred to is that at a position angle of about 15° to the sun Galactic Centre line of sight (see, for example, Dwek et al. 1995) and is in reality a triaxial bulge. However, this cannot be causing the features at $l=21°$ and $27°$. If the features at $l=27°$ and $l=21°$ are associated with a bar then it is distinct from the triaxial bulge and would have to have a position angle of about $75°$. Hence the Galaxy would in effect be a double barred spiral.

5. Conclusions

New JH star counts are presented for various directions on the Galactic plane looking towards the inner disc. The analysis of the disc counts shows that the extinction in the region of the molecular ring ($R=4.5$ kpc) is some three times higher than would be expected from a simple extrapolation from the solar circle.

The colour-magnitude diagrams show the presence at $l=27°$ $21°$ and $15°$ of a major giant branch which is not present at $l=31°$. This giant branch accounts for some 50% of the detected H sources at $l=27°$ and the feature causing the branch is at a distance of about 7.5 kpc. A bar provides the simplest explanation for the distribution in the inner disc of both the old stars detected here and young stars detected in the TMGS. A ring could be made to fit the data but the distribution of sources would be *ad hoc* and patchy. If the bar does turn out to be the explanation, then the Galaxy is a two barred spiral.

References

Clemens D. P., Sanders D. B., Scoville N.Z. 1988, ApJ,327, 139
Dwek E et al. 1995, ApJ. 445, 716.
Garzón F., Hammersley P. L., Mahoney T., Calbet X., Selby M. J., Hepburn, I., 1993, MNRAS, 264,773
Hammersley P.L., Garzón F., Mahoney T. ,Calbet X. 1994, MNRAS, 269 753
Kent S.M., Dame T. M., Fazio, G., 1991, ApJ 378 131.
Sodroski.T.J., Odegard,N., Arendt R.G., Dwek E., Weiland J.L., Hauser M.G., Kelsali T., 1997, ApJ 480, 173

SPATIAL DISTRIBUTION OF EVOLVED GIANT STARS IN THE GALACTIC DISC USING DENIS DATA

STÉPHANIE RUPHY
Observatoire de Paris, DESPA, 92190 Meudon, France
stephanie.ruphy@larecherche.fr

Abstract. I will present the analysis of 3-colour (IJK$_s$) near infrared star counts obtained during the commissioning phase of the Deep Near Infrared Southern Sky Survey (DENIS) project. Nine strips of 12' of RA ×30° in declination crossing the galactic plane at different longitudes ranging from 217 °to 385 °, covering approximately 50 square degrees are analyzed. One strip has been measured in 3 colours, the others in J and K only. More than 250 000 objects are detected and calibrated. The completeness limits are 17.5, 15.5 and 13.5 in the I, J and K$_s$ bands, respectively. Colour-colour diagrams are shown to be efficient tool for breaking out dwarf and giant star populations especiall y at low latitude, where the interstellar extinction strongly affects the far away red giants. Source counts are compared to the SKY model developed by Cohen (1997). This small sample of DENIS data is tentatively used to investigate the existence and the spatial distribution of the stellar populations associated with the molecular ring at R = 4 kpc. The detail of my work has been published in Ruphy et al., (1996), Ruphy, (1996) and Ruphy et al., (1997).

References

Cohen M. (1996) Proc. of the 2nd Euroconference on *The impact of Large Scale near-infrared sky surveys*, Puerto de la Cruz, Spain, eds. F. Garzon, N. Epchtein, A. Omont, W.B. Burton, P. Persi, Kluwer ASSL series **vol. no. 210**, pp 67–70

Ruphy S. (1996) *Thèse de Doctorat, Université de Paris 6*

Ruphy S., Robin A.C., Epchtein N., Copet E., Bertin E., Fouqué P., Guglielmo F. (1996) *A&AL*, **Vol. no. 313**, pp. L21-L24

Ruphy S., Epchtein N., Copet E., de Batz B., Borsenberger J., Fouqué P., Kimeswenger S., Lacombe F., Le Bertre T., Rouan D., Tiphène D. (1997) *A&A*, **vol.326**, pp. 597-607

N. Epchtein (ed.),
The Impact of Near-Infrared Sky Surveys on Galactic and Extragalactic Astronomy, 69.
© 1998 *Kluwer Academic Publishers.*

III- Stars

III. Stew

INFRARED EMISSION OF CARBON STARS

M. A. T. GROENEWEGEN
Max-Planck-Institut für Astrophysik,
Karl-Schwarzschild-Straße 1, D-85748 Garching, Germany

Abstract.
This review addresses the infrared emission from the most important classes of carbon-rich objects: the N- and R-type carbon stars, the carbon dwarfs and the CH-stars. Examples of carbon rich objects in the Galactic halo and the Magellanic Clouds are discussed. Discrimination of carbon-rich objects against red oxygen-rich objects is investigated. The detectibility of N-type carbon Miras in the DENIS and 2MASS near-infrared surveys is discussed.

1. Introduction

The organisers had asked me to review the infrared emission of carbon stars. They may not have realised that this is a formidable task indeed as there are actually many types of carbon-rich objects. Here, I will not discuss the R CrB stars, hydrogen deficient carbon (HdC) stars and carbon-rich post-AGB stars. The latter two classes have no or very little present-day mass loss and their near-infrared emission (most important for the kind of infrared surveys that are the topic of this conference) is therefore dominated by the stellar photosphere. Instead, I will concentrate on the more well-known classes of carbon-rich objects, namely the N- and R-type carbon stars, the CH-stars and the carbon dwarfs. In particular, I would like to discuss the spectral characteristics of these classes of carbon-stars.

As will be discussed below, the near-infrared DENIS and 2MASS surveys have the potential to discover carbon stars with near-infrared emission to large distances. In addition, the related question how to distinguish the different types of carbon-rich objects amongst each other and from red oxygen-rich objects from the data of these surveys will be investigated using colour-colour diagrams.

N. Epchtein (ed.),
The Impact of Near-Infrared Sky Surveys on Galactic and Extragalactic Astronomy, 73-86.
© 1998 *Kluwer Academic Publishers.*

2. Different kinds of carbon stars

2.1. THE N-TYPE CARBON STARS

The N-type carbon stars are formed on the Asymptotic Giant Branch (AGB) after one or more third dredge-up events where carbon is mixed into the convective envelope. For a comprehensive scenario see Groenewegen et al. (1995), for a general review on AGB stars, see Habing (1996).

Characteristic of this class of stars is their variability and heavy mass loss, which are related to each other. As is the case for oxygen-rich Miras, there are correlations between the pulsation period and luminosity (Feast et al. 1989, Groenewegen & Whitelock 1996), and period and mass loss rate (Groenewegen et al. 1997b) for carbon miras.

2.2. THE R-TYPE CARBON STARS

The R-type carbon stars are hotter and less luminous than the N-type (Scalo 1976) and can not be on the AGB. These, or very similar hot and low luminosity carbon stars, have been found in the Galactic bulge (Lloyd Evans 1985, Westerlund et al. 1991) and the Magellanic Clouds (Westerlund et al. 1992, 1995).

McClure (1997) monitored 22 R-stars and found no evidence for radial velocity variations in any of them.

The classical explanation for the formation of these carbon stars considers an extraordinary mixing event at the Helium core flash (e.g. Dominy 1984). Based on the lower than expected 20% binary frequency McClure (1997) proposes that these stars may be coalescing binaries.

Significant cold and hot dust is known in only one case (Parthasarathy 1991). For the general absence of significant IR excess in R-, CH- and Ba II-stars see Dominy (1986).

2.3. THE CH-TYPE CARBON STARS

These stars, recognised as a separate class by Keenan (1942), are typically metal-poor objects with the kinematics of Population II objects. They are in binary systems (McClure & Woodsworth 1990). They are hotter than the N-type, with equivalent spectral types in the range G-K. Their near-infrared colours are of photospheric origin.

2.4. THE CARBON DWARFS

Currently, there are nine of these peculiar objects known. Their space density exceeds that of N-type carbon stars (Green et al. 1992). The prototype carbon dwarf G77-61 was discovered twenty years ago (Dahn et al. 1977) and long remained the only example known. Then, eight more were discovered by Green et al. (1991, 1992), Heber et al. (1993), Warren et al. (1993) and Liebert et al. (1994). Three carbon dwarfa are known binaries; this was established for the prototype from radial velocity measurements, for the two others from composite spectra. In the latter cases the companion is a white dwarf. The absolute V-magnitude is in the range +9.7-10.8.

3. Carbon stars in different environment

3.1. HALO CARBON STARS

Carbon stars in the Galactic halo have been long known (Sanduleak 1980, Margon et al. 1984). They are sometimes designated faint high-latitude carbon (FHLC) stars. Margon et al. (1984), Mould et al. (1985), Bothun et al. (1991), Green et al. (1992) and Moody et al. (1997) contain in total 41 such stars, some of which turned out to be dwarf carbon stars (Green & Margon 1994), or are known CH-stars. Originally, the interest in them was as tracers of the outer halo.

In a recent paper, Totten & Irwin (1997) present the results of an optical survey (a byproduct of the APM survey) for FHLC stars covering about 6500 sq. degrees. Thirty-two are definitely N-type, and some of them may be 'dusty'.

Interestingly enough, some FHLC stars are losing mass at a considerable rate. Two were serendipitously discovered by Cutri et al. (1989) and Beichman et al. (1990), namely IRAS 08546+1732 and IRAS 12560+1656. Both are Miras with a 390 day period (Joyce et al. 1997). A more detailed analysis is performed by Groenewegen et al. (1997a). From the period-luminosity relation of Groenewegen & Whitelock (1996) a luminosity of 5800 $L_\odot$ is derived for a period of 390 days, measured for both stars. A model fit with a dust radiative transfer model to the spectral energy distributions then gives the distances of 8.0 and 20 kpc from sun, and 7.8 and 11 kpc from the Galactic plane.

CO J = 2-1 emission was detected in the nearest object (IRAS 12560) after ten hours of integration using the IRAM 30m telescope. The expansion velocity is low at only 3.2 km s^{-1}, and stellar LSR velocity is +88 km s^{-1}. Combining the CO and dust observations and the modelling, Groenewegen et al. (1997a) derive for this star a mass loss rate of 1.3 × 10^{-6} $M_\odot$yr^{-1}, and a dust-to-gas ratio of 0.0014. The abundance of CO w.r.t H_2 is found to be 2 × 10^{-4}, indicating an underabundance of oxygen by –0.7 dex. w.r.t. solar.

The colours of these two stars are as follows: M_K = −6.8 and −7.6, I_C = 14.3 and 18.7, J = 10.4 and 14.0, H = 8.6 and 11.8, K = 7.0 and 9.7. With such colours and a K-survey down to 14th magnitude one could probe the

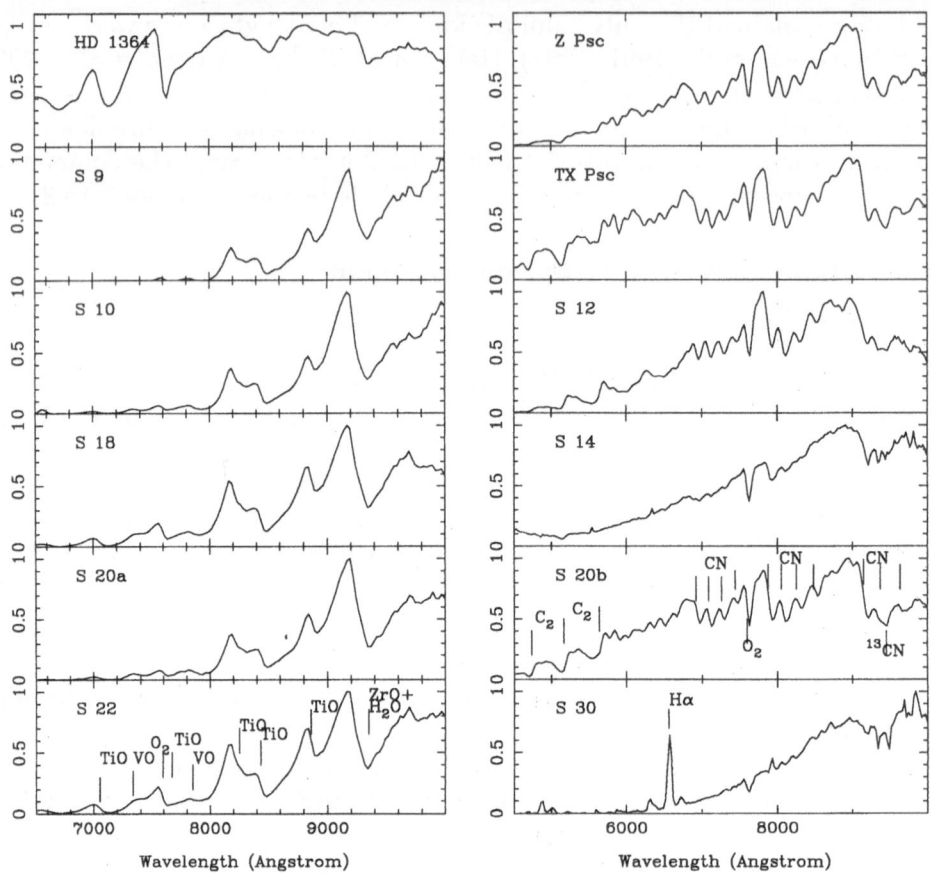

Figure 1. Optical spectra of AGB stars in the SMC (identified by S <number>), the M3.5 IIIa MK standard star HD 1364 and the known carbon stars TX Psc and Z Psc. Well known lines are identified, including telluric O_2 and H_2O lines. From Groenewegen & Blommaert (1997).

halo out to distances of at least 150 kpc. It is expected that 2MASS and DENIS will identify many more mass-losing halo carbon stars.

3.2. CARBON STARS IN THE MAGELLANIC CLOUDS

This is a long standing research topic dating back to the first surveys in the 1960's by Westerlund (1960). I had the opportunity to review this subject recently, including a discussion of the first ISO results on AGB stars in the Magellanic Clouds (Groenewegen 1997b). I will therefore be brief on this topic. Just as a reference: the total estimated number of carbon stars in the LMC is about 11 000 (Blanco & McCarthy 1983), while about 2 900 are known in the SMC (Blanco & McCarthy 1983, Rebeirot et al. 1993).

The last few years saw a number of interesting papers (WEG: taking a look at this subject) from the point of view of IRAS detected AGB stars in the Clouds (Whitelock et al. 1989, Reid et al. 1990, Reid 1991, Wood et al. 1992, Groenewegen et al. 1995, Zijlstra et al. 1996, van Loon et al. 1997a, b, Loup et al. 1997, Groenewegen & Blommaert 1997). The first and last paper in this list specifically deal with the SMC, the other papers concentrate on the LMC. The important outcome of these papers is that mass-losing AGB stars exist in the Clouds. However, as the 12 μm fluxes are close to the detection limit, the IRAS view is far from complete. For a given mass loss rate, the IRAS detection probability increases with increasing luminosity, and vice versa. An almost complete census of AGB stars in the Clouds will be provided by DENIS and 2MASS.

4. Spectral characteristics of carbon stars

In this section I will describe the emission and absorption features one can observe in carbon stars, starting in the optical and then turning to the near- and mid-infrared.

In the optical the absorption bands that identify a carbon star are the C_2 bands at e.g. 4737, 5165 and 5636 Å, and the stronger CN bands longwards of 7000 Å. As an example, Fig. 1 shows low resolution optical spectra of carbon (and oxygen-rich AGB stars for comparison) stars in the SMC, together with some well known objects (Groenewegen & Blommaert 1997).

A comprehensive atlas of carbon star spectra can be found in Barnbaum et al. (1996; at moderate resolution) and Barnbaum (1994; at high resolution).

Although one could have thought that the optical spectra of carbon stars are well understood almost a century since their discovery, the recent paper by Sarre et al. (1996) provides the first accurate vibrational band assignments of the SiC_2 molecule. This molecule has prominent bands in the 4750-5050 Å region (also called Merrill-Sanford bands). Figure 2 is an adaptation of Fig. 1 of Sarre et al. (1996) where details can be found. See Barnbaum et al. (1996) for more examples.

Green et al. (1992) find unusually strong C_2 6191 Å bandheads in the carbon dwarfs, which is not seen in any of the other FHLC stars they observed, and suggest this may be a spectroscopic way of identifying carbon dwarfs.

Moving towards the near-infrared there are two prominent absorption bands that identify carbon stars: a C_2 band at 1.77 μm at the edge of the H-band, and a blend of HCN and C_2H_2 lines (see Ridgway et al. 1978 for a high resolution spectrum) near 3.1 μm at the edge of the L-band.

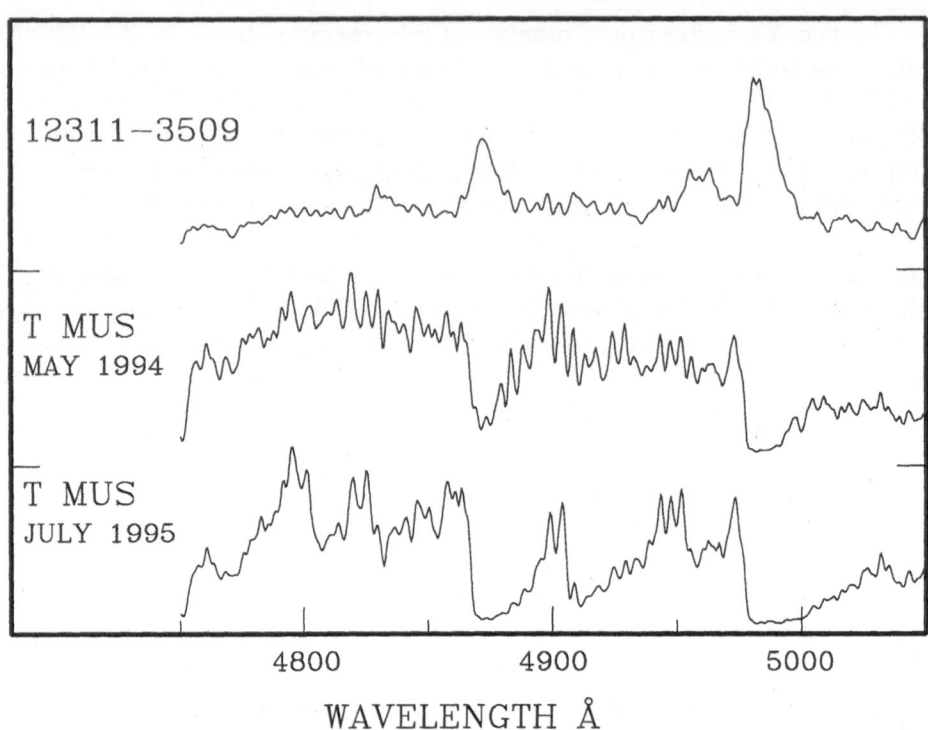

WAVELENGTH Å

Figure 2.   Spectra between 4750 and 5050Å of IRAS 12311-3509 and the semiregular variable T Mus in May 1994, when the SiC$_2$ absorption spectrum was abnormal, and July 1995. The spectra are flux-calibrated on an arbitrary intensity scale. The absorption bands at 4872, 4911 and 4982 Å are clearly visible in the bottom spectra. The spectrum of 12311–3509 illustrates that these bands can also be observed in emission. Adopted from Sarre et al. (1996), where more details can be found.

An example of the 1.77 μm feature is shown in Fig. 3 (from Lançon & Wood 1997). More near-infrared spectra of carbon stars are shown in Lázaro et al. (1994).

An example of the shape of the 3.1 μm feature is shown in Fig. 4 (from Groenewegen et al. 1994; also see Yamamura et al. 1997b). Substructure is clearly visible due to the fact that this feature is a blend. More 3 μm spectra of carbon- and oxygen-rich stars can be found in Merrill & Stein (1976a,b,c).

Recently, the Japanese IRTS and ESA's ISO satellite provided a wealth of new information, because of the coverage of wavelengths not observable from the ground, and the higher spectral resolution. IRTS/NIRS-MIRS spectra of carbon stars are shown in Fig. 5 (adopted from Yamamura et al. 1997a). ISO SWS spectra are shown in Fig. 6 (adopted from Yamamura et al. 1997b). The following lines can be identified: HCN + C$_2$H$_2$ features

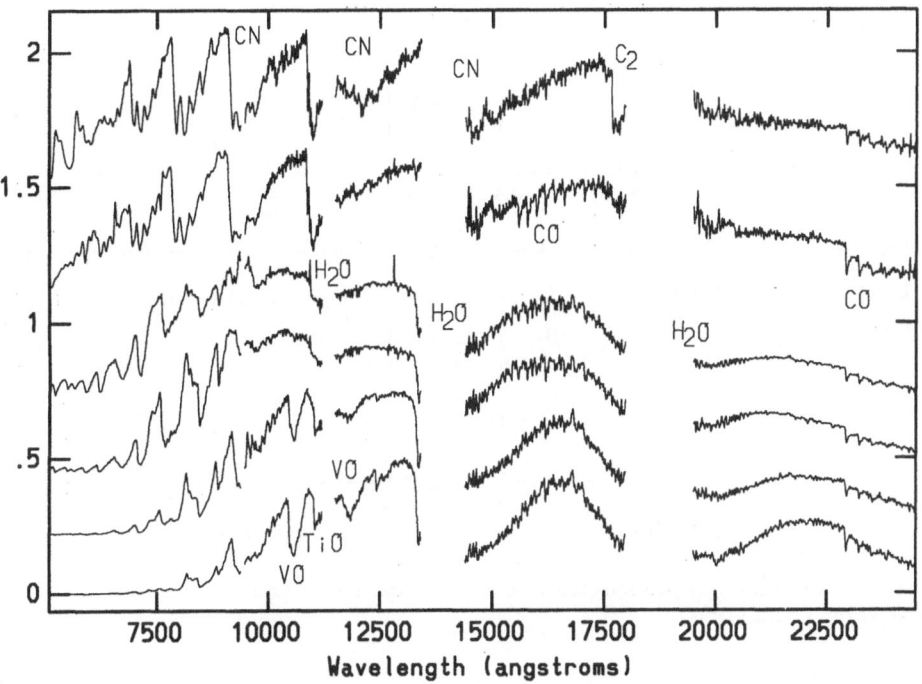

Figure 3. Near-infrared spectra of a carbon star (top spectrum) compared to oxygen-rich spectra. Note the strong C_2 absorption at 1.77 μm. From Lançon & Wood (1997).

at 3.1 μm (see above), 3.9 μm (also observable from the ground, e.g. Bregman et al. 1978), 7.5 μm and 13.7 μm, a C_3 feature at 5.2 μm, and the well-known silicon carbide (SiC) dust emission feature at 11.3 μm. I will not discuss features longward of 20 μm. The SiC feature can be observed from the ground (e.g. Treffers & Cohen 1974, Speck et al. 1997) and is present in the wavelength range covered by the IRAS low resolution spectrograph, and is thus studied extensively (e.g. Lorentz-Martins & Lefèvre 1994, Groenewegen et al. 1997b)

The SWS spectra have revealed so much detail that the theoretical interpretation is lacking behind the observations at the moment. Recent attempts to remedy this, and e.g. to predict the strength of the various absorption features as a function of phase in the pulsation cycle, can be found in Loidl et al. (1997) and Hron et al. (1997).

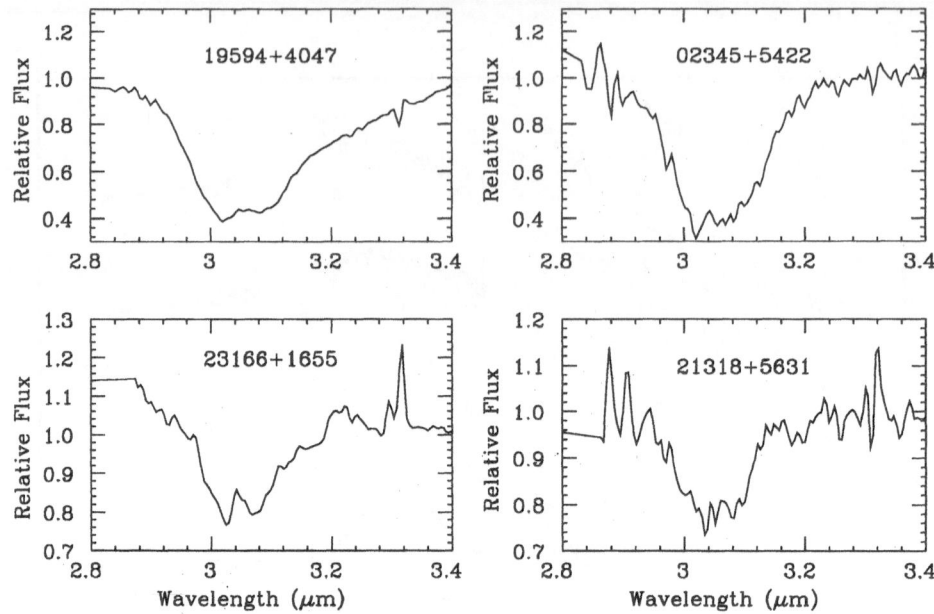

Figure 4.   Normalised 3 μm spectra of carbon stars. IRAS names are indicated. From Groenewegen et al. (1994).

5. Carbon stars in relation to DENIS and 2MASS

Let us first estimate to which distances carbon Miras can be detected in the DENIS and 2MASS surveys. As the K-band seems the most suitable filter in terms of sensitivity and intrinsic brightness of the objects in question, I concentrate on that filter. In Groenewegen & Whitelock (1996) the following period-K-relation is derived from stars in the LMC, scaled to the galaxy: $M_K = -3.56 \log P + 1.14$.

In Table 1 the relevant numbers are collected for a typical 'short' period of 200 days, an 'average' period of 400 days and a typical 'long' period of 700 days. The respective absolute K-magnitudes based on the P-K-relation are listed, as well as the range in distances carbon miras can be observed based on saturation and 10σ detection limits for DENIS (Epchtein 1997) and 2MASS (Skrutskie et al. 1997). The conclusion is that the near-IR surveys have the potential to detect carbon stars out to several hundred kpc.

Can the different types of carbon-rich objects be distinguished from each other, and from oxygen-rich objects?

Figure 7 shows IJK and JHK colour-colour diagrams from Groenewegen (1997a). Oxygen-rich and carbon-rich AGB stars are compared to

IRTS/NIRS-MIRS Spectra of Carbon Stars

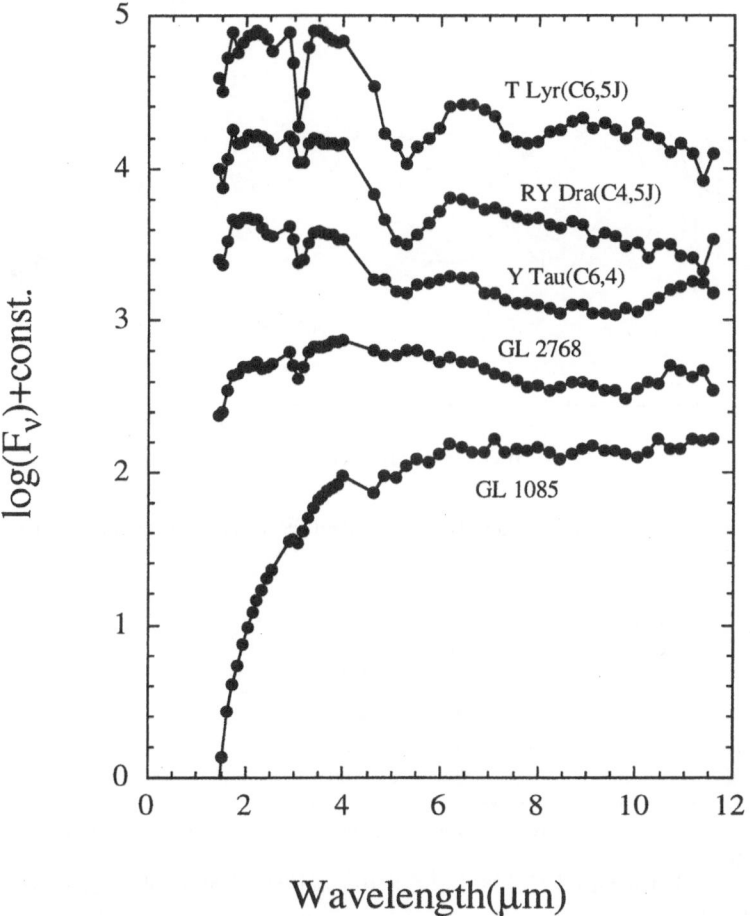

Figure 5. IRTS spectra of carbon stars. Note the strong features at 3.1 μm and 5.2 μm. Adopted from Yamamura et al. (1997a).

galaxies, M-dwarfs, Herbig Ae/Be stars and T Tau stars. For the construction of these samples see Groenewegen (1997a). Note that the sequence for the oxygen-rich AGB stars (the filled squares) does not appear continuous. This is an artifact of the type of templates used.

Additional information comes from Green et al. (1992, their Fig. 1) who present a colour-colour diagram which includes the carbon dwarfs. It appears that in a $(J - H), (H - K)$ diagram they are off-set with respect to other FHLC-stars and the low luminosity carbon stars in the Galactic

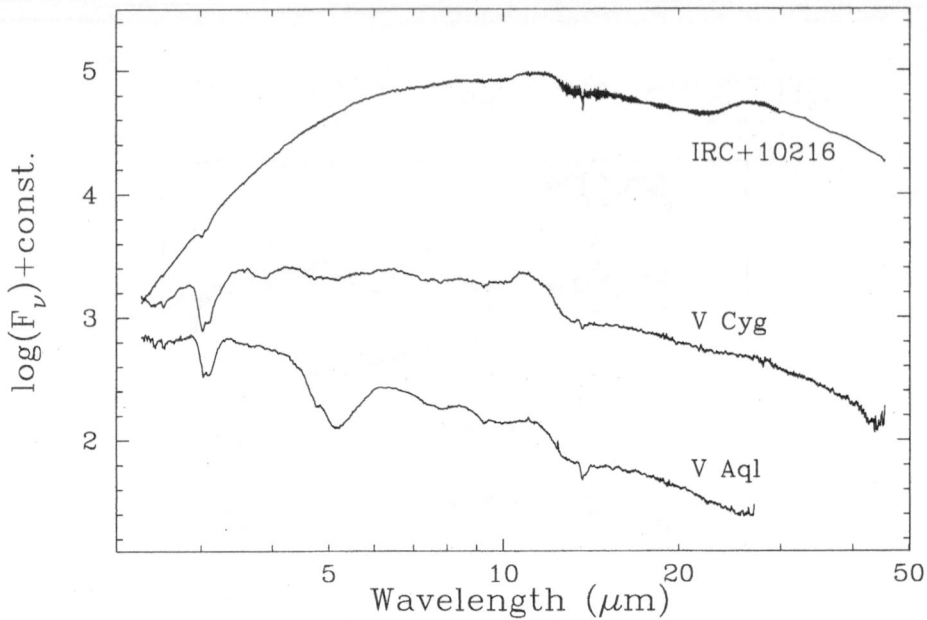

Figure 6.   ISO SWS spectra of carbon stars. Adopted from Yamamura et al. (1997b).

TABLE 1.   How far can carbon miras be detected ?

	survey	$P = 200$	400	700 days
M_K		−7.05	−8.12	−8.95
Range	DENIS[a]	5.1-90	8.4-145	12.5-220
(kpc)	2MASS[b]	1.6-190	2.7-300	3.9-450

[a]Based on a saturation limit of $K = 6.5$ and a 10σ detection limit of 12.7. Interstellar extinction is neglected.
[b]Based on a saturation limit of $K = 4.0$ and a 10σ detection limit of 14.3. Interstellar extinction is neglected.

bulge. The colours of the carbon dwarfs are blue: $0.25 \lesssim (H - K) \lesssim 0.36$, $0.35 \lesssim (J - H) \lesssim 0.65$.

Furthermore, Feast & Whitelock (1992, their Fig. 5) present a $(J - H), (H - K)$ colour-colour diagram that includes CH-stars in the LMC, carbon dwarfs in dwarf spheroidals, Galactic CH- or related stars, CH-stars in ω Cen and the dwarf carbon stars. Apart from the dwarf carbon stars that stand out as mentioned above, there is a large overlap between the classes. They also show that there is a significant overlap with normal oxygen-rich giants.

For comparison, Fig. 8 shows colour-colour diagrams for a relatively

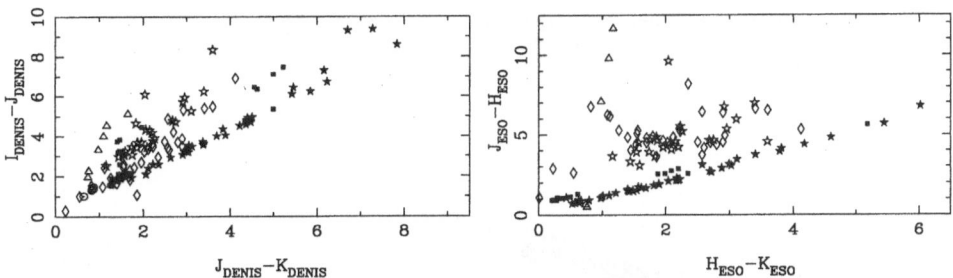

Figure 7. *IJK* and *JHK* colour-colour diagram. Symbols: carbon stars = filled stars; oxygen-rich AGB stars = filled squares; galaxies = open circles; M-dwarfs = open triangles; He Ae/Be stars = open diamonds; T Tau stars = open stars. The limits of the *IJK* plot are set based on the saturation and 5σ detection limits of DENIS; very red AGB objects outside the plot limits do exist. For the AGB stars the colours are dereddened. From Groenewegen (1997a).

unbiased sample of carbon stars in the LMC (Costa & Frogel 1996). The larger scatter in the bottom panel is in all likelihood due to the fact that the *I*- and *JK* photometry was not taken simultaneously.

The conclusion I draw from all these data is that, based on *IJK* and *JHK* colour-colour diagrams alone, (a) it is extremely difficult to separate unreddened carbon stars (mainly CH-stars, R-stars) from normal oxygen-rich giants, as exception possibly being the carbon dwarfs, (b) for moderate red colours there is the danger of overlap with other red sources. This danger appears to be less for the 2MASS system, (c) mass-losing AGB stars with very red colours are easily identified but this will be a minority of sources amongst the 10^8 that are expected to be detected, as demonstrated in Fig. 7.

As discussed in Groenewegen (1997a) additional information may be used to make a more clear-cut discrimination, e.g. optical colors or IRAS fluxes. However the latter only provides such information for a few 10^5 sources only.

6. Conclusions

Carbon stars display a rich absorption and emission spectrum. The recent satellite missions ISO and IRTS have unveiled parts of the spectrum that are inaccessible from the ground, the 5-8 μm and the $\gtrsim 20$ μm region.

The upcoming near-infrared surveys have the potential to detect carbon stars out to large distances. However, except for the reddest sources, it will prove difficult to separate the different classes of carbon stars from colour-colour-diagrams alone and there is the potential of confusion with oxygen-rich sources, or other red sources. Conclusive identification could be done using narrow-band filters in the optical (e.g. Richer et al. 1984)

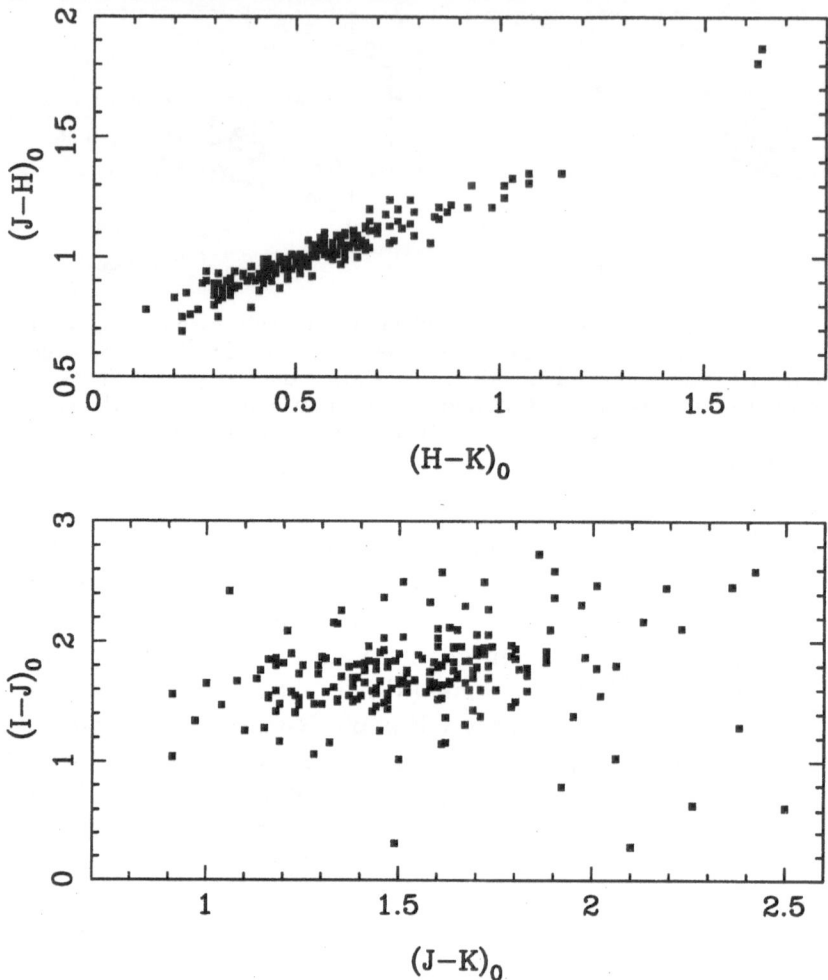

Figure 8.   Colour-colour diagram for a sample of carbon stars in the LMC. Based on data from Costa & Frogel (1996).

or the near-infrared (not yet attempted). This should be possible even for distant objects with the new generation of 10m class telescopes.

ACKNOWLEDGEMENTS

I would like to thank the people who granted me permission to include (modified) version of published figures and for their effort in making these figures: P. Sarre, T. Lloyd Evans, J. Blommaert, A. Lançon, I. Yamamura, and their collaborators.

References

Barnbaum C., 1994, ApJS 90, 317

Barnbaum C., Stone R.S., Keenan P.C., 1996, ApJS 105, 419

Beichman C.A., et al., 1990, AJ 99, 1569

Blanco V.M., McCarthy M.F., 1983, AJ 88, 1442

Bothun G., et al., 1991, AJ 101, 2220

Bregman J.D., Goebel J.H., Strecker D.W., 1978, ApJ 223, L45

Costa E., Frogel J.A, 1996, AJ 112, 2607

Cutri R.M., et al., 1989, AJ 97, 866

Dahn C.C., Liebert J., Kron R.G., Spinrad H., Hintzen P.M., 1977, ApJ 216, 757

Dominy J.F., 1984, ApJS 55, 27

Dominy J.F., 1986, AJ 91, 951

Epchtein N., 1997, in: "The impact of large-scale IR surveys", eds. F. Garzón et al., p. 15

Feast M.W., Glass I.S., Whitelock P.A., Catchpole R.M., 1989, MNRAS 241, 375

Feast M.W., Whitelock P.A., 1992, MNRAS 259, 6

Green P.J., Margon B., 1994, ApJ 423, 723

Green P.J., Margon B., Anderson S.F., MacConnell D.J., 1992, ApJ 400, 659

Green P.J., Margon B., MacConnell D.J., 1991, ApJ 380, L31

Groenewegen M.A.T., 1997a, in: "The impact of large-scale IR surveys", eds. F. Garzón et al., p. 165

Groenewegen M.A.T., 1997b, in: "ISO's view on stellar evolution", eds. L. Waters, C. Waelkens and K. A. van der Hucht, Kluwer Academic Publishers, p. TBD

Groenewegen M.A.T., Blommaert J.A.D.L., 1997, A&A, submitted

Groenewegen M.A.T., van den Hoek L.B., de Jong T., 1995, A&A 293, 381

Groenewegen M.A.T., de Jong T., Geballe T.R., 1994, A&A 287, 163

Groenewegen M.A.T., Oudmaijer R.D., Ludwig H.-G., 1997a, MNRAS in press

Groenewegen M.A.T., Smith C.H., Wood P.R., Omont A., Fujiyoshi T., 1995, ApJ 449, L119

Groenewegen M.A.T., Whitelock P.A., 1996, MNRAS 281, 1347

Groenewegen M.A.T., Whitelock P.A., Smith C.H., Kerschbaum F., 1997b, MNRAS in press

Habing H., 1996, A&AR 7, 97

Heber U., Bade N., Jordan S., Voges W., 1993, A&A 267, L31

Hron J., Loidl R., Jørgensen U.G., Kerschbaum F., in: "ISO's view on stellar evolution", eds. L. Waters, C. Waelkens and K. A. van der Hucht, Kluwer Academic Publishers, p. TBD

Joyce R.R., Merrill K.C., Gillett F.C., 1997, in preparation

Keenan P.C., 1942, ApJ 96, 101

Lançon A., Wood P.R., 1997, in : "Poster proceedings of IAU Symp. 189 on Fundamental stellar properties: The interaction between Observation and Theory", p.168, ed. T.R. Bedding, School of Physics, University of Sydney

Lázaro C., et al., 1994, MNRAS 269, 365

Liebert J., et al., 1994, ApJ 421, 733

Loidl R., Höfner S., Hron J., Aringer B., Jørgensen U.G., in: "ISO's view on stellar evolution", eds. L. Waters, C. Waelkens and K. A. van der Hucht, Kluwer Academic Publishers, p. TBD

Lloyd Evans T., 1985, MNRAS 216, 29P

van Loon J. Th., Zijlstra A.A., Whitelock P.A., Waters L.B.F.M., Loup C., Trams N.R., 1997a, A&A 325, 585

van Loon J. Th., Zijlstra A.A., Whitelock P.A., te Lintel Hekkert P., Chapman J.M., Loup C., Groenewegen M.A.T., Waters L.B.F.M., Trams N.R., 1997b, A&A in press

Loup C., Zijlstra A.A., Waters L.B.F.M., Groenewegen M.A.T., 1997, A&AS, in press

Lorentz-Martins S., Lefèvre J., 1994, A&A 291, 831

Margon B., Aaronson M., Liebert J., 1984, AJ 89, 274
McClure R.D., 1997, PASP 109, 256
McClure R.D., Woodsworth A.W., 1990, ApJ 352, 709
Merrill K.M., Stein W.A., 1976a, PASP 88, 285
Merrill K.M., Stein W.A., 1976b, PASP 88, 294
Merrill K.M., Stein W.A., 1976c, PASP 88, 874
Moody J.W., Gregory S.A., Soukup M.S., Jaderlund E.C., 1997, AJ 113, 1022
Mould J.R., Schneider P., Gordon G.A., Aaronson M., Liebert J., 1985, PASP 97, 130
Parthasarathy M., 1991, A&A 247, 429
Rebeirot E., Azzopardi M., Westerlund B.E., 1993, A&AS 97, 603
Reid N., 1991, ApJ 382, 143
Reid N., Hughes S.M.G., Glass I.S., 1995, MNRAS 275, 331
Richer H.B., Crabtree D.R., Pritchet C.J., 1984, ApJ 287, 138
Ridgway S.T., Carbon D.F., Hall D.N., 1978, ApJ 225, 138
Sanduleak N., 1980, PASP 92, 246
Sarre P.J., Hurst M.E., Lloyd Evans T., 1996, ApJ 471, L107
Scalo J.M., 1976, ApJ 206, 474
Skrutskie M.F., et al., 1997, in: "The impact of large-scale IR surveys", eds. F. Garzón
    et al., p. 25
Speck A.K., Barlow M.J., Skinner C.J., 1997, MNRAS 288, 431
Treffers R., Cohen M., 1974, ApJ 188, 545
Totten E.J., Irwin M.J., 1997b, MNRAS, submitted
Warren S.J., Irwin M.J., Evans D.W., Liebert J., Osmer P.S., Hewett P.C., 1993, MNRAS
    261, 185
Westerlund B.E., 1960, Uppsala Astron. Obs. Ann. 4, No. 7
Westerlund B.E., Azzopardi M., Breysacher J., Rebeirot E., 1992, A&A 260, L4
Westerlund B.E., Azzopardi M., Breysacher J., Rebeirot E., 1995, A&A 303, 107
Westerlund B.E., Lequeux J., Azzopardi M., Rebeirot E., 1991, A&A 244, 367
Whitelock P.A., Feast M.W., Menzies J.W., Catchpole R.M., 1989, MNRAS 238, 769
Wood P.R., Whiteoack J.B., Hughes S.M.G., Bessell M.S., Gardner F.F., Hyland A.R.,
    1992, ApJ 397, 552
Yamamura, I. and the IRTS team, 1997a, in "Infrared Diffuse Radiation and the IRTS",
    eds. H. Okuda, T. Matsumoto, and T. L. Roellig, PASP 124, 72
Yamamura I., de Jong T., Justtanont K., Cami J., Waters L.B.F.M., 1997b, in: "ISO's
    view on stellar evolution", eds. L. Waters, C. Waelkens and K. A. van der Hucht,
    Kluwer Academic Publishers, p. TBD
Zijlstra A.A., Loup C., Waters L.B.F.M., Whitelock P.A., van Loon J. Th., Guglielmo
    F., 1996, A&A 279, 32

AGB STARS IN THE GALACTIC BULGE
OBSERVED BY DENIS

M. SCHULTHEIS
Institut für Astronomie der Universität Wien, Austria

G. SIMON
Observatoire de Paris, France

AND

J.HRON
Institut für Astronomie der Universität Wien, Austria

Abstract. We present first results of DENIS photometry for semiregular variables (SRVs) and Miras in field #3 of the Palomar-Groningen survey (PG3, $l = 0^0, b = -10^0$). The PG3 Miras and SRVs are located in the colour-colour diagram (CCD) in both colours at the reddest end. PG3 variables show a large scatter in $(I - J)_0$ while in a $K_0^S/(J - K^S)_0$ diagram they are situated at the top of the red giant branch. In contrast to the LMC we do not find any carbon star sequence which is due to the different age and metallicity of the Bulge. PG3 variables follow a PC relation as well as a period-luminosity relation.

1. Introduction

PG3 (field #3 of the Palomar-Groningen Variable Star Survey; $l = 0^0$, $b = -10^0$) is well searched for variable stars (Plaut 1971, Wesselink 1987). Blommaert (1992) studied the properties of the PG3 Miras while Schultheis et al. (1997) the properties of the SRVs. Besides near-Infrared photometry (JHKL′M) obtained at the 1 m ESO telescope (now dedicated to DENIS), spectra and radial velocities are available. Schultheis et al. (1997) found that the PG3 SRVs are not the analogs of the field SRVs. The PG3 SRVs form a short period extension to the Miras PK and PC relations. This indicates that the PG3 Miras and SRVs are both pulsating in the same

87

N. Epchtein (ed.),
The Impact of Near-Infrared Sky Surveys on Galactic and Extragalactic Astronomy, 87-94.

mode, possibly the fundamental. The metallicity of PG3 is between half solar and solar. Both PG3 Miras and SRVs follow the SgrI PK relation which again stresses the metallicity independence of the PL relation. Ng & Schultheis (1997) found for a few faint AGB stars evidence that they are members of the Sagittarius dwarf galaxy. Two carbon stars found in PG3 might be member of the Sagittarius darf galaxy as well (Ng, 1997).

The main aim of this work is to obtain DENIS photometry for those well-known AGB stars. Their position in a colour-colour and a colour-magnitude diagram will help to search for more AGB stars candidates. Further on, DENIS will offer a good tool to study the stellar population related to the Sagittarius dwarf galaxy.

2. Selection of DENIS strips

Based on the coordinates of the PG3 variables we looked for 77 Miras and 78 SRVs in the DENIS archive. 21 PG3 Miras and 15 PG3 SRVs have been observed by DENIS up to now. At least two measurements for 6 Miras and 2 SRVs were obtained. Fig. 1 shows the observed DENIS strips of PG3. In

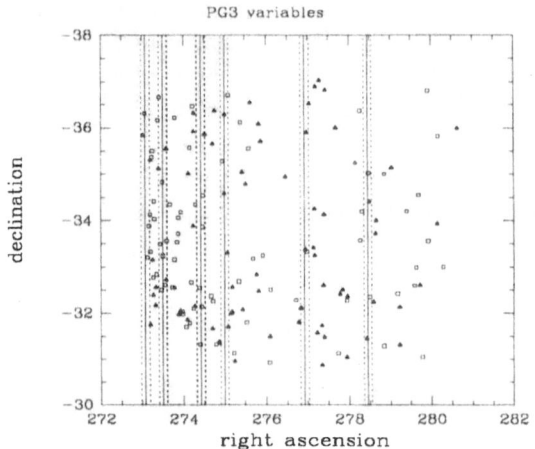

Figure 1.   Distribution of the PG3 variables in α and δ. PG3 SRVs are indicated by triangles, PG3 Miras by squares. Solid lines mark the centre of the observed DENIS strips, the dashed ones indicate each 6'.

total 8 strips were observed by DENIS where two of them were reobserved (see table 1), which gives a total number of $\sim$ 80 frames. For two strips the quality is very poor and have to be reobserved. Note that we do not deal with full strips but with single frames.

strip number	Date obs.	α	comments
2801	26/08/95	18 33 50	I channel not available
2887	14/09/95	18 20 00	
4099	19/05/96	18 17 41	
4109	22/05/96	18 27 41	
4423	14/08/96	18 12 18	
4438	16/08/96	18 13 04	
4485	22/08/96	18 33 50	different epoch as 2801
4721	11/10/96	18 20 00	different epoch as 2887

TABLE 1. Observed DENIS strips for PG3

3. Reduction procedure

The whole reduction procedure was done with the PDAC pipeline. For each strip two standard stars were used in order to derive the photometric zero point. In the subsections below the reduction procedure is described roughly. For a more detailed description I refer to Ruphy (1996).

3.1. ABSOLUTE ASTROMETRY

For the absolute astrometry the individual DENIS frames were cross-correlated with the Guide Star Catalog in each channel. As we are dealing with crowded fields the automatic correlation failed and it had to be done manually. The corresponding offset has been applied to the frames. Afterwards for each frame the distortion matrix was calculated. The absolute astrometry is accurate in the order of ~ 0.5".

3.2. SOURCE EXTRACTION

The Sextractor software written by E.Bertin was used for extracting the sources. For a description of the Sextractor itself I refer to Bertin and Arnouts (1996). In the remaining catalog, bad sources have been removed. Near the edge of the frame some sources are not extracted to avoid border problems. Fig. 2 demonstrates the difference in the extraction procedure as well as the systematic shift in the astrometry. LDAC extracts in total more sources than we found, using the whole frame and extracting deeper in magnitude (due to different cuts used).

PG3 field: M. Schultheis (1997)

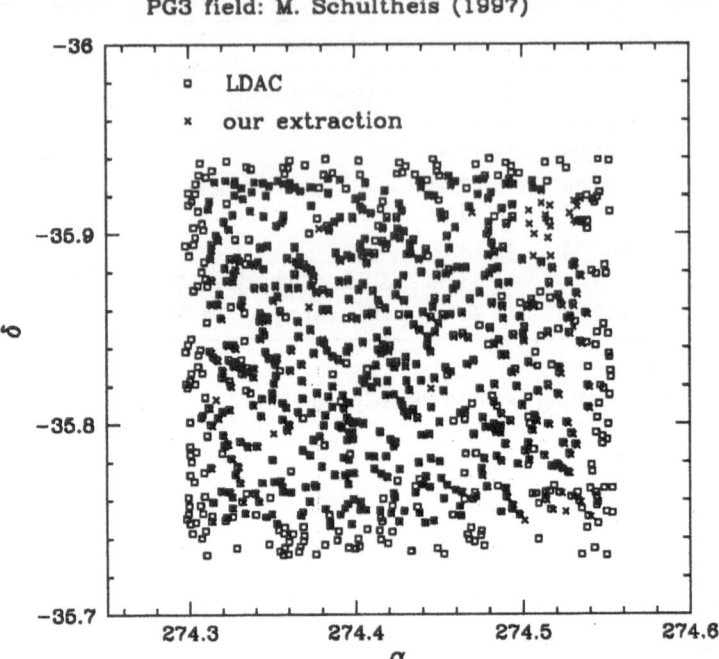

Figure 2. Comparison of the LDAC and our source extraction for frame Nr. 531977. The sources are detected in all three channels. Open squares indicate the LDAC source extraction, crosses our source extraction.

3.3. PHOTOMETRY

As we used only two standard stars for the determination of the zero point a standard extinction coefficient has been applied. For some strips the standard star was clearly saturated, mostly in the I channel, so that only one star was used for the definition of the zero point. The standard deviation of the 8 single measurements taken for the standard PHOTO is typically of the order of a few hundreths of a magnitude. The mean over the 8 values was taken for the zero point. Between the different strips the zero point can deviate up to $0\overset{m}{.}8$ which is most likely due to different atmospheric conditions for the various strips.

3.4. PAIRING

The pairing procedure determines the reliability of the colours one gets and is therefore very important, especially for crowded fields where misidentifications lead to wrong colours. In our case the pairing was done between the three channels for each frame separately. In an iterative procedure the best match between all channels has been obtained. The resulting accuracy of the pairing is of the order of 0.1".

3.5. IDENTIFICATION OF THE PG3 VARIABLES

The PG3 variables were identified mainly based on their coordinates and their known magnitudes in J and K. In a few cases, there were a few possible candidates very close to the position. In this case finding charts helped to identify the correct object. Especially, some PG3 Miras are saturated and could not be used for further analysis.

3.6. QUALITY ANALYSIS

For each strip separately a colour-colour diagram was obtained and compared with each other. For some strips the colour-colour diagrams look very broad and show a significantly large offset which is mostly due to the quality of the night. Those strips have been excluded. They further show in contrast to the others a very broad magnitude distribution. In total four strips were taken into account for further analysis. Between the four strips there is a noticeable shift in the magnitude distribution. We shifted the remaining strips with regard to the reference strip. The magnitude distribution observed here comes from the gaussian peaked shape of the bulge population superimposed to the non-gaussian distribution of the disk population. The faint FWHM is approximately the place where the combined field stars and bulge RGB contribution ought to become dominant. Since this is not the case we assume that this hints for a large incompleteness in the stellar sample or the detection limit of the survey due to increased crowding. Our detection limits are therefore $I = 14^{m}.8$, $J = 14^{m}.3$ and $K^{S} = 13^{m}.8$ respectively. We only regarded sources where the photometric error in each channel is less than $0^{m}.1$. All sources have been corrected for interstellar extinction according to Wesselink (1987).

4. Results

Fig.3 shows the $(I - J)_0/(J - K^S)_0$ diagram. The AGB stars cover in both colours the reddest end. In $(I-J)_0$ the colour range is very large and goes up until $4^m.0$. Especially the AGB stars show a large variation of $(I-J)_0$. Loup (see her contribution at this conference) shows very nicely two branches in the colour-colour diagram for the LMC.

One is related to the oxygen-rich stars while the other correspond to the c-rich objects. Her result is very well in agreement with our fig. 3 where we clearly miss the c-rich branch. The absence of carbon stars in the Bulge is due to a different age and metallicity compared to the LMC.

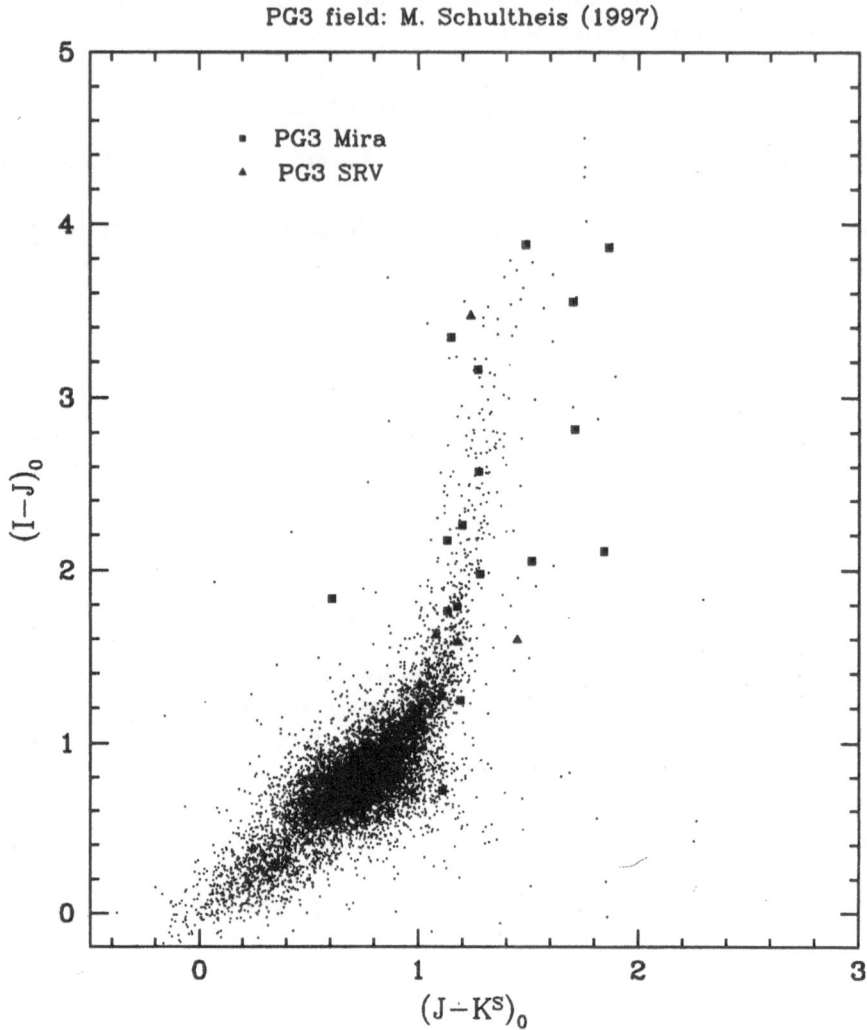

Figure 3. CCD of PG3 stars. All stars have been corrected for interstellar extinction according to Wesselink (1987). PG3 SRVs are indicated by triangles, PG3 Miras by squares.

In a $K_0^S/(J - K^S)_0$ diagram one can clearly see the tip of the horizontal branch, the red giant branch and the AGB (see fig. 4). The PG3 Miras and SRVs are situated at the tip of the red giant branch. A closer look shows that the AGB is separated from the RGB. This separation is not very clear due to the variability of those stars. The upper limit of the scatter in J–K due to variability is ~ 0.2 mag for Miras and ~ 0.1 mag for SRVs. The variation in K is about 1 mag for Miras and 0.5 mag for SRVs (Hron &

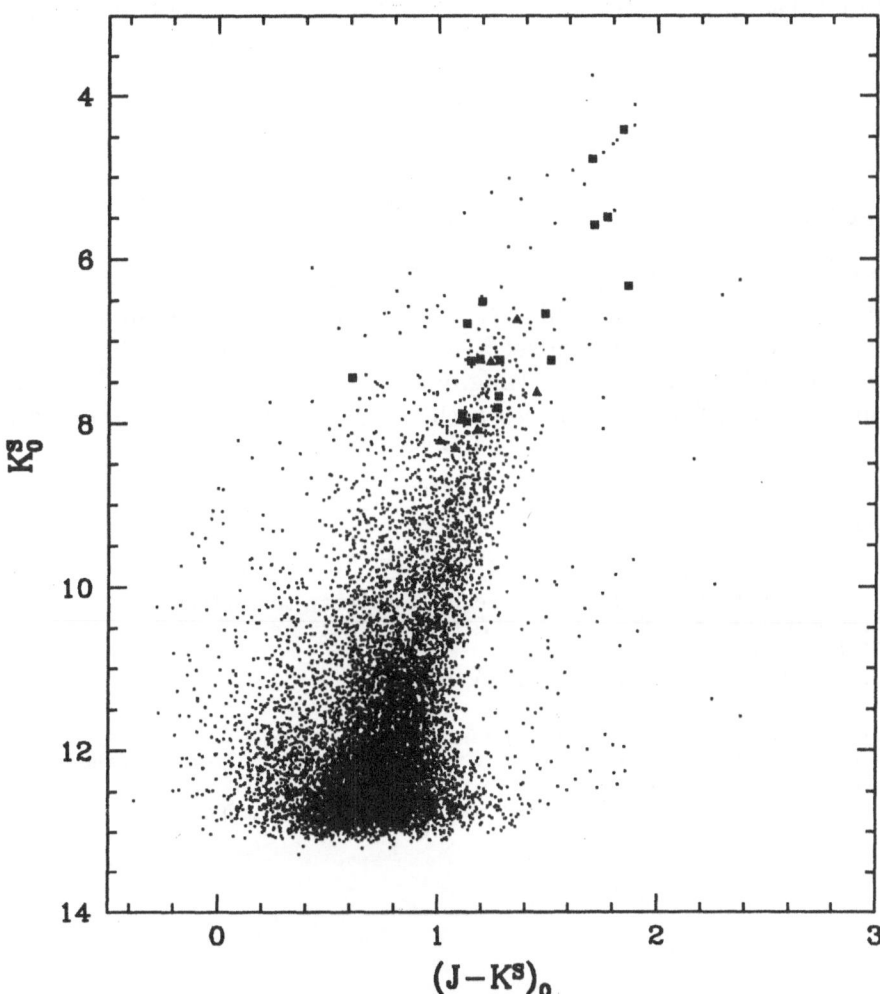

Figure 4. CMD of PG3 stars. The symbols are the same as fig. 3. All stars have been corrected for interstellar extinction according to Wesselink (1987).The limiting magnitudes are I = $14^{m}.8$, J = $14^{m}.3$ and K^{S} = $13^{m}.8$ and $\sigma_I < 0^{m}.1$, $\sigma_J < 0^{m}.1$ and $\sigma_K < 0^{m}.1$

Kerschbaum 1994). Again in comparison to the LMC we miss the branch of the carbon stars. The PG3 variables follow a PK_0^S relation. The slope of the PK_0^S relation is the same as the PK relation. As a first approach we determined the shift between K and K^S assuming that the slope remains the same. We obtained a $\Delta(K - K^S)$ of ~ 0.4. Further on PG3 variables follow in J–K, I–J and I–K a period-colour relation.

5. Outlook

DENIS CMDs and CCDs will enable us to find new AGB star candidates. Isochrones from the Padova group (e. g. Bertelli et al. 1994) will allow us to derive metallicities and ages. Further on with DENIS we will be able to study in detail the stellar populations related to the Sagittarius dwarf galaxy.

Acknowledgements

The authors thank Y. K. Ng, A. Omont, C. Loup, M. R. Cioni and D. Ojha for constructive suggestions. M. Schultheis thanks the Institut d' Astrophysique where part of this research was carried out, for their hospitality. We are especially grateful to F. Tanguy and J. Borsenberger for their help in using the DENIS data. The research of M. Schultheis is supported by the Austrian Science Fund projects P9638-AST and S7308.

References

Bertin, E., Arnouts, S., (1996) A&AS Sup. 117, 393
Blommaert, J.A.D.L (1992), phD thesis, Leiden University, the Netherlands
Hron J., Kerschbaum F., (1994), ApSS 217, 137
Ng, Y.K. (1997), A&A, in press
Ng, Y.K., Schultheis, M. (1997), A&AS 123, 115
Plaut, L., A&AS 4, 75
Ruphy, S. (1996), phD. thesis, University Paris 6
Wesselink, Th. J. H (1987), Ph. D. thesis, Catholic University of Nijmegen, the Netherlands.
Willems,F.J, (1987), Ph. D thesis, University of Amsterdam, the Netherlands

ANALYSIS OF DENIS DATA IN ISOGAL FIELDS

C. ALARD

DASGAL, 77 avenue Denfert Rochereau, Observatoire de Paris.
and Institut d'Astrophysique de Paris.

1. Introduction

We present here DENIS color magnitude diagrams for 3 different fields in
the Milky Way. One field is situated in a window at only 2 degrees from
the Galactic Center, the two others are in the Galactic plane at longitudes
l=15 and l=25. We first show that it is possible to probe the extinction law
along the line of sight using K vs. (J-K) color magnitude diagrams, and
we also demonstrate that the shape of the giant branch is changing with
Galactic latitude. For these fields ISO data from the ISOGAL project are
also available. These ISOCAM data have been reprocessed using the CIA
package (see acknowledgements), allowing reconstruction of a high quality mosaic image from the raw ISOCAM data. Magnitude and positions of
stars are extracted from the ISOCAM mosaic image and associated to DE-
NIS K band point sources. These DENIS K magnitudes are an important
asset to our ISO 7 and 15 micron fluxes, especially in the case of AGB stars.

2. Probing the extinction law along the line of sight.

The field close to the Galactic Center, is a new window closer to the Galactic Center than previous studies in Baades's Window (Tiede *et al.* 1995).
But the extinction in this field is too low to probe with some efficiency
the extinction law. However the color magnitude diagrams for the two others fields located in the Galactic plane at longitudes l=15, and l=25 (see
Fig. 1) indicate a significant reddening, quite variable with distance. Three
sequences are visible in the diagrams(from left to right): a main sequence,
the sequence traced by the giant clump associated with this main sequence,
and finally an upper giant branch from stars in the central region of the
Galaxy. The Second sequence, the clump of giant stars is especially useful
because its absolute magnitude and unreddened colors lie within a narrow
range. Consequently, measuring the colors and magnitudes of this clump
will give immediatly the extinction law. Looking back at the diagrams, we

95

N. Epchtein (ed.),
The Impact of Near-Infrared Sky Surveys on Galactic and Extragalactic Astronomy, 95-100.
© 1998 *Kluwer Academic Publishers.*

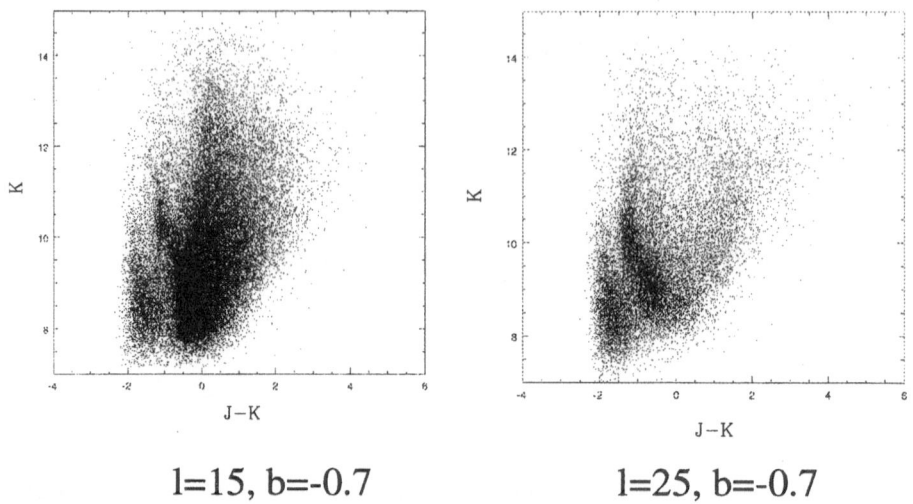

Figure 1. The variations of extinction with distance. Note the sudden change in the foreground giant clump color. Magnitudes are not calibrated, zero points remain to be estimated.

see that the color variation with distance of the clump is very similar for the two diagrams. A sudden variation of the clump color occurs at a distance that we estimate to be about 3 Kpc (we wait until making a real absolute calibration to be more specific). It means that a large amount of extinction, and consequently of interstellar material is concentrated at this distance. This feature could be associated with the Sagittarius arm, but might also be related to the molecular ring. At this time, it is difficult to guess what the right interpretation is. Although, further investigations with more DENIS data, and especially a selection of new fields evenly sampled from the center to a longitude of about 45 deg. would be very helpful. The new ISOGAL/DENIS reductions which will be achieved soon will certainly provide us with a suitable material.

3. Changes in the giant branch shape with Galactic longitude.

The DENIS I,J,K data it is possible to derive a magnitude and a color which are independent of reddening. We have only to assume a reddening law (Frogel and Withford, 1987) to derive the correction to reddening, we derive: $K' = K - 0.538 \times (J - K)$ and, $(I - J)' = (I - J) \times 2.15(J - K)$. In figure 2 we plot the K' vs. (I-J)' diagrams for our three fields. We immediately notice a tail on the right of the diagram, for bright magnitudes in the Galactic center diagram. This tail is not found in the two other diagrams. Note also that the upper part of the diagram is more and more depopu-

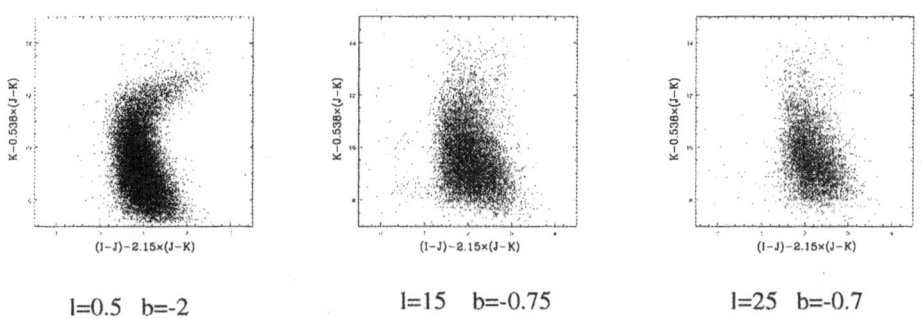

l=0.5 b=-2   l=15   b=-0.75   l=25 b=-0.7

Figure 2. We plot here the color/magnitude diagrams corrected for extinction in 3 different Galactic fields. Note the change with latitute observed in the region of the diagram situated at bright magnitude and red colors. Magnitudes are not calibrated, zero points remain to be estimated.

lated with increasing longitude. This feature, which exists only close to the Galactic Center shows that a particular population is present in the central region. It is tempting to relate these giants with redder colors to a metal rich population, which we expect to find in the Galactic Bulge. Mapping the extension of this population by counting the stars in the region occupied by the feature would provide us with an interesting insight concerning the structure of the Galactic Bulge. Such mapping is essential in order to understand the formation and chemical evolution of our Galaxy. It would require to get DENIS data for all the central region.

4. The ISO data.

The ISOGAL project is a survey of the Galactic plane and central regions of the Milky Way. We made observations at 7 and 15 microns for a few hundreds fields evenly sampled along the Milky Way. The ISOCAM infrared camera of the satellite makes a series of 32x32 images for each field. Each $32 times 32$ small image is repated about 15 to 10 times in order to create a time series, and from the whole stack of images a mosaic image of the field is constructed. The image reconstruction process is not simple, and require different corrections, the most difficult beeing the transient correction. These transients are due to the detector memory effect, if an area of the detector is illuminated with a source, the relevant pixels will not drop to zero immediately after the source is off. This memory effect will create ghost images of a previously seen source, which will take long to disappear. This is especially annoying for our crowded Galactic fields, and is the main limitation to the depth of our images. An illustration of this memory effect

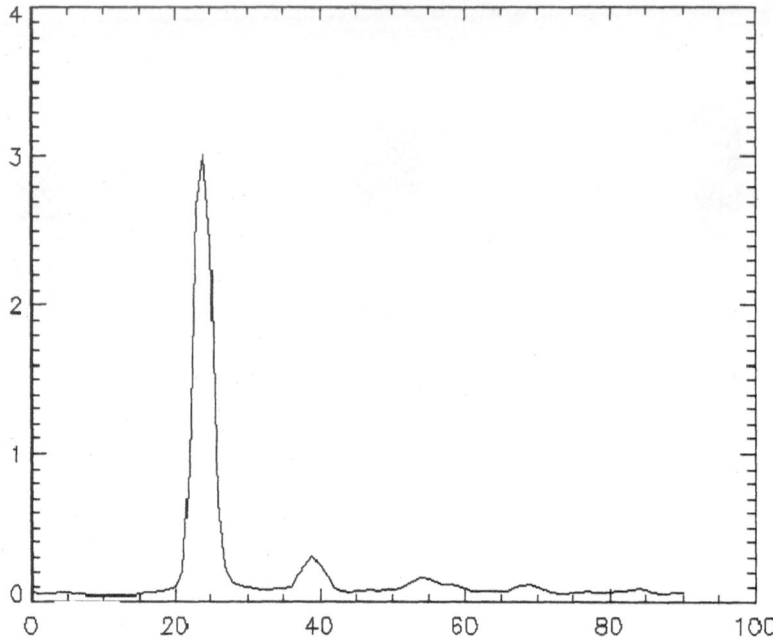

Figure 3. Memory effect of ISOCAM. The detector sees first a bright source on the left of the diagram. Note the decreasing ghost images of this bright source visible as a periodic sequence.

is given in Fig. 3. Another important issue is that this difficult processing is not completely satisfactory in the standard ISO/ESA pipeline. The data had to be reprocessed using the CIA package, and in particular the J.L. Starck wavelet method was used for correction of the memory effects. A dramatic improvement of the point source photometry and repeatability was achieved using CIA. This improvement is also obvious just by looking at the images (see Fig. 4).

4.1. CROSS IDENTIFICATION WITH DENIS SOURCES

Each ISO source is searched for a DENIS source within a small radius around the ISO source. We start with a radius of 10 arcsec, which allows us to find a first list of candidate associations. A differential fit is then performed, and a new search for DENIS sources is conducted. We iterate the process until convergence. Our typical resulting rms error in the ISO/DENIS associations is about 1.5 arcsec. We perform the final associations at 3 σ, which gives an eventual association radius of 4.5 arcsec. The resulting diagram is given in Fig. 5. for a field near the Galactic Center. Note that the density of DENIS sources is so high that it is difficult to avoid false associations like the clump of faint stars on the right of the diagram. However, the likelihood of associations with bright DENIS sources is much

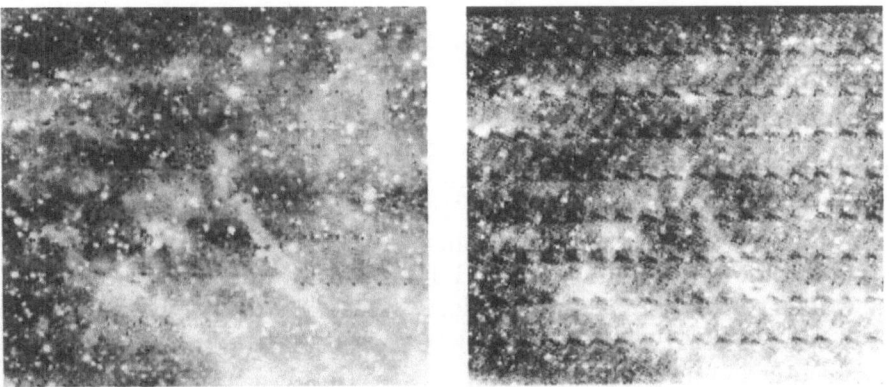

Figure 4. Comparison of data processing methods. Left is the original mosaic image from the ISO/ESA pipeline, and right is the newly processed image, using the CIA package.

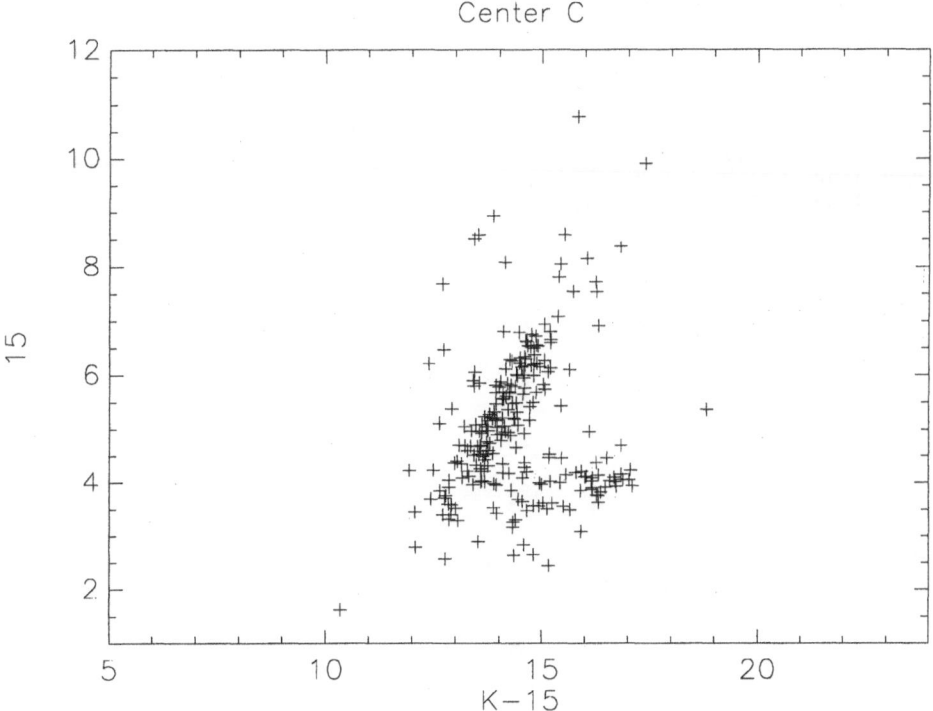

Figure 5. The ISO vs. DENIS associations for a field situated near the Galactic Center. See text for discussion.

higher, and the sequence on the left of the diagram is real. This sequence is associated with bright giants, most of them beeing probably AGB stars. We estimate that about 70 % of ISO sources are found in real associations with DENIS sources.

5. Conclusion.

We demonstrate here that interesting results can be derived from the DE-NIS observations of the ISOGAL fields. It would be certainly interesting to pursue these investigations with DENIS data covering a more extended region around the Bulge. We also show that these DENIS data can be well completed with ISO data at 7 and 15 micron. Processing of ISO data is now efficient and reliable, and we demonstrate that most of ISO source are retrieved in DENIS with excellent positions and magnitude correlation.

Acknowledgements

The ISOCAM data presented in this paper was analysed using "CIA", a joint development by the ESA Astrophysics Division and the ISOCAM Consortium led by the ISOCAM PI, C. Cesarsky, Direction des Sciences de la Matiere, C.E.A., France.

References

Alard, C. 1997, in preparation
Frogel, J. and Withford, A.E., 1987, ApJ, 320,199
Tiede, G., et al., 1995, AJ, 110, 2788

IN QUEST OF VERY LOW MASS STARS AND BROWN DWARFS WITH NEAR-IR SURVEYS

T. FORVEILLE, X. DELFOSSE
Observatoire de Grenoble
Université de Genoble, France

AND

N. EPCHTEIN
Observatoire de la Côte d'Azur, France

1. Introduction

Deep large scale near-infrared surveys have started revolutionizing the study of very low mass stars and brown dwarfs (hereafter VLMS and BD) by producing, for the first time, statistically significant samples. They represent the most efficient observational method to discover the so called *free-floating* BDs that for decades have been looked for, as well as large unbiased samples of VLMS of spectral types later than M6.5V, whose mass and luminosity functions are still poorly known and understood.

The efficiency of deep near–IR surveys simply results from their probing the 1-2 micron spectral window, where objects with effective temperature in the 1500-3000K range radiate the bulk of their bolometric luminosity. Based on a black-body reasoning, red and brown dwarfs would be expected to have extreme infrared colours for any bandpass combination, but the deep molecular absorption bands that characterize these very cool stars mean that this is not always true. For well chosen filter combinations however, like the I−J colour provided by DENIS, red and brown dwarfs do show up in uncrowded areas of near IR colour–colour diagrams. The only other stellar populations that exhibit similar colours are the late giants (red giants, AGBs stars, extreme carbon stars, etc..). As they are more luminous by 5 to 10 orders of magnitudes, and have a rather flat galactic distribution, this confusion is not difficult to remove at least outside a 40 degrees band around the galactic equator. There is on the other hand no obvious way, to distinguish between BDs and VLMS on the sole basis of the infrared colours, except for the very coolest BDs. Spectroscopic follow up observation is

101

N. Epchtein (ed.),
The Impact of Near-Infrared Sky Surveys on Galactic and Extragalactic Astronomy, 101-108.
© 1998 *Kluwer Academic Publishers.*

thus an essential step to definitely confirm a colour selected candidate as a genuine brown dwarf. The so-called *Lithium test*, which consists in detecting the 6807 Å absorption of Li in the atmosphere of the star, is at present the most conclusive confirmation of a candidate as a *bona fide* brown dwarf (Martín et al., 1995).

The two ongoing large scale near infrared surveys, are likely to produce a wealth of exciting data on BDs. The discovery of a sample of good candidates and the confirmation of at least one BD in a small sample of DENIS data is extremely encouraging for the future. Hundreds of BDs are likely to be effectively singled out in the coming years. They will for the first time yield a statistically significant set of such stars and the first direct measurement of the low-end of the solar neighbourhood luminosity function. The parallel intense modeling activity (e.g., Tsuji, Allard et al., 1997, Chabrier and Baraffe, 1997) will undoubtedly also provide breakthroughs in this exciting area of astrophysics.

2. The quest for brown dwarfs

Not so long ago, brown dwarfs were still considered as hypothetical objects (Tinney, 1995) although stellar evolution theories have long predicted the existence of astronomical objects whose mass spans the range between a few Jupiter masses (hereafter M_J) and some 70 M_J, and whose luminosity is smaller than $10^{-4} L_\odot$. They fill the mass and luminosity gaps that separate giant planets orbiting around a star (such as Jupiter, about 1/1000th of solar mass) and low mass stars (of a few hundredth of solar mass). They are not massive enough to initiate the combustion of hydrogen in their core, hence their chemical composition has not varied much along their evolution. For this simple reason, one of the most efficient test to confirm an object as a brown dwarf is the Li test. Objects with masses lower than some $60 M_J$ must have kept their Li since the very efficient proton collision reaction $[\mathrm{Li}^7(\mathrm{p}, \alpha)\mathrm{He}^4]$ easily destroys the Li in some 10^7 yrs, at temperatures lower than are needed to initiate hydrogen burning.

As brown dwarfs have extremely low luminosities, all early searches used indirect methods and looked for the reflex astrometric or radial velocity motion induced on a brighter binary companion star. All of these early attempts actually failed to single out any serious candidate (e.g., Stevenson, 1991), but recent radial velocity searches with much improved technology have isolated several unseen brown dwarf companions to bright nearby stars (Mayor, private communication). The quest for brown dwarf was revived in the mid 1980s by the development of panoramic IR detectors and high spatial resolution technics such as speckle interferometry and coronography, which have to a large extent taken over the indirect detection methods.

Since that time, several authors have claimed to discover genuine brown dwarf candidates. They almost invariably turned out to be VLMS rather than BDs, with only GD165B (Becklin & Zuckermann, 1988) remaining as an undecided borderline case.

2.1. CONFIRMED BROWN DWARFS: PRESENT STATUS

A good overview of the early and recent searches for BDs has been recently published by Kulkarni (1997). Here we will only briefly summarize the status of presently known BDs and update his compilation with the most recent results. Known BDs can be observationally split into two groups, young BDs (younger than roughly 0.3 Gyear) found near their formation site, and old BDs (older than roughly 1 Gyear) found in the field or around a bright nearby star.

2.1.1. *Young BD's,*

Young BDs are characterised by much higher luminosities (by several orders of magnitude) than old BDs, because they haven't yet had time to cool down. They can consequently be searched for at much larger distances, in nearby star forming regions and stellar associations. Searches in regions of recent star formation such as the Taurus or Ophiuchus clouds have essentially failed, mainly because the strong dust extinction that characterize these regions hampers accurate luminosity determination. In addition the Li test is inadequate there, since *all* low mass stars retain a large fraction of their Li for the first 10^7 years of their life. For this reason, BD searches have now concentrated on the nearby Pleiades open cluster, whose age is $\sim 10^8$ years. A systematic deep CCD survey (with an 80 cm telescope !), followed up by high resolution spectroscopy on the Keck 10m telescope resulted in the discovery of the first confirmed young BD (Teide 1), with an estimated mass of 55 $\pm$15 M_J (Rebolo et al., 1995; Rebolo et al., 1996). This has now been followed by the detection of a wealth of new BDs and VLMS (Zapatero-Osorio et al., 1997).

2.1.2. *Old BD's*

Gliese 229B, was the first confirmed BD and was found during a search for low-mass companions to nearby stars ($\leq 15pc$) (Nakajima et al., 1995). This program used a specially designed coronographic instrument on the 60" telescope at Palomar Observatory. Gl 229B remains, by a large margin, the BD with the coolest known effective temperature (~ 1200 K). Its near-infrared spectrum is characterized by extremely strong CH_4 absorption bands, and is in many respects intermediate between those of warmer BDs and of the giant planets.

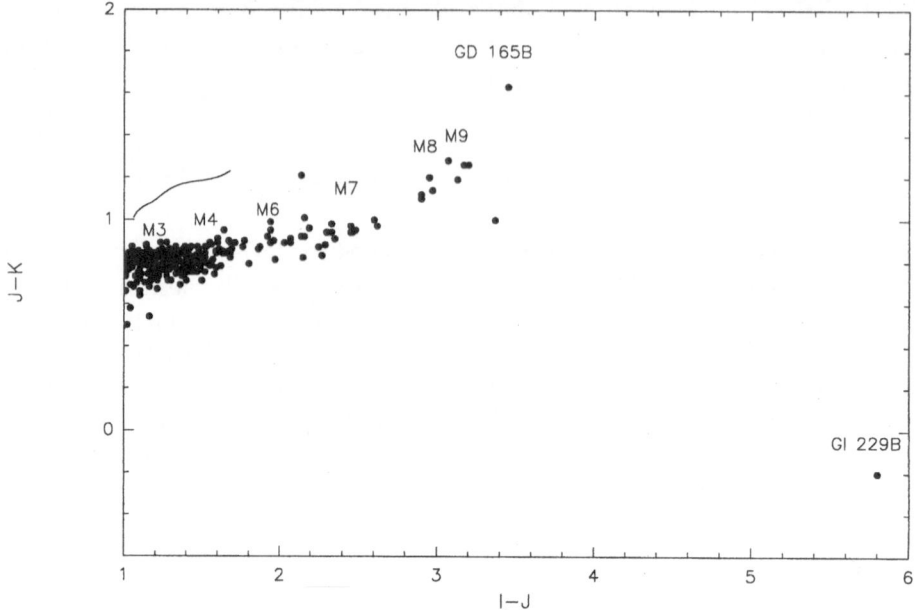

Figure 1. The I–J:J–K colour-colour diagram for previously known very low mass-stars.
The short continuous line shows the giant sequence.

2.2. THE DENIS BROWN DWARFS AND VERY LOW MASS STARS

The set of DENIS colours (IJK) is particularly well suited to identify VLMS
and BD (Reid, 1994). As illustrated in Figure 1, the [I-J] colour has large
variations over the relevant spectral type range, when [J-K] (and [J-H],
not shown here) have a much smaller dynamical range, and even become
bluer beyond the spectral type of GD 165B (Gl 229B has the same J-
K colour as a B star). It is thus relatively easy to single out BDs and
VLMS even from relatively noisy measurements near the DENIS sensitivity
limit, while from JHK measurements this would only be possible, if at all,
from accurate photometry of well detected sources. For such a program the
DENIS sensitivities at I and J are in addition quite well matched, as they
cross-over at spectral type M6V-M7V.

Using some of the first DENIS data, a program aimed at selecting extreme red dwarfs, the *"DENIS brown dwarf mini-survey"*, was carried out
(Delfosse, 1997; Delfosse et al., 1997). The analysed area consists of a set
of 52 strips totalizing 230 square degrees. Their distribution on the sky is
largely random, except that their distance from the galactic plane is always
greater than 20 ^{o} to avoid crowding and contamination by red giants, AGB
stars, and distant reddened objects. Image processing was performed in the
standard DENIS pipeline (Borsenberger, 1996), but source extraction was

done using SExtractor package. A 3 colour catalog of objects brighter than
I = 18.5, J = 16 and K = 13.5 was produced, in which some 50 sources
have [I-J] ≥ 2.5.

Follow-up near IR spectroscopy of all objects redder than [I-J] ≥ 2.8 was
then obtained at AAT using the IRIS spectrograph. This confirmed that
the three reddest objects, namely DENIS-P J0205.4-1159, 1058.7-1548 and
1228.2-1547, were indeed good BD candidates. Further optical spectroscopic
observations were then performed at the Keck telescope (Martin et al.,
1997) and at the AAT (Tinney et al., 1997) for last two objects. They show
a spectacular LiI absorption line in 1228.2-1547, but not in 1058.7-1548.
DENIS-P J1228.2-1547 thus became the first *bona-fide* isolated (or free-
floating) BD, together with Kelu 1, which was simultaneously discovered
by Ruiz et al. (1997) during a proper motion survey aimed at cool white
dwarfs. Another isolated BD candidate, called 296A, has also passed the
Li test (Thackrah et al., 1997), but its status should nonetheless probably
still be regarded as uncertain: if it is indeed a brown dwarf, its relatively
high effective temperatures implies an uncomfortably young age for a field
object.

The three DENIS BD candidates, Kelu-1, and GD 165B have optical
spectra which are unlike those of the late M dwarfs, without the strong TiO
and VO bands that characterize the M spectral type. This is presumably
due to condensation of these refractory molecules onto dust grains (Allard,
1997), and leads Martín et al. (1997) to suggest that these spectra should
be attributed a new spectral class, for which they propose L.

TABLE 1. Properties of confirmed brown dwarfs. *(Notes in col. 6, 1: Nakajima
et al. 1995; 2: Rebolo et al., 1996; 3: Ruiz et al., 1997; 4: Delfosse et al., 1997;
5: Tinney et al., 1997; 6: Thackrah et al., 1997)*

Name	Type	d (pc)	Mass (MJ)	Teff (K)	Réf.
Gl229B	Companion	5.7	≤ 50	≤1200	1
Teide 1	Pleiades	125	55 ± 15	2600±150	2
Calar 3	Pleiades	125	55 ±15	2600 ± 150	2
Kelu 1	isolated	12	≤ 75	1900 ± 100	3
DENIS-P J1228-1547	isolated	?	≤ 60	1600 ± 300	4,5
296A	isolated	?	65 ± 25	2800 ± 200	6

TABLE 2. M dwarfs later than spectral type M7V in de DENIS minisurvey (from Delfosse, 1997).

Name	I	I–J	J–K	I–K
DENIS-P 0909 − 0658	17.21 ±0.08	3.20 ±0.09	1.50 ±0.09	4.70 ±0.12
DENIS-P 0912 − 0414	16.71 ±0.06	2.46 ±0.07	1.10 ±0.12	3.56 ±0.13
DENIS-P 0910 + 0019	18.40 ±0.20	2.65 ±0.26	-	-
DENIS-P 0944 − 1305	16.72 ±0.05	2.59 ±0.06	0.76 ±0.17	3.35 ±0.18
DENIS-P 0940 − 2257	17.02 ±0.10	2.79 ±0.11	0.99 ±0.16	3.78 ±0.19
DENIS-P 0944 − 0900	18.30 ±0.18	2.69 ±0.24	-	-
DENIS-P 0944 − 1310	18.50 ±0.21	2.80 ±0.24	-	-
DENIS-P 1007 − 1706	17.73 ±0.14	2.84 ±0.16	-	-
DENIS-P 1154 + 0135	16.03 ±0.04	2.77 ±0.04	1.09 ±0.06	3.86 ±0.07
DENIS-P 1227 + 0114	16.96 ±0.10	2.88 ±0.11	1.29 ±0.15	4.17 ±0.17
DENIS-P 1228 − 1547	18.19 ±0.27	3.76 ±0.27	1.70 ±0.16	5.46 ±0.31
DENIS-P 1228 − 2415	18.00 ±0.20	3.11 ±0.21	-	-
DENIS-P 1058 − 1548	17.80 ±0.17	3.72 ±0.17	1.37 ±0.15	5.09 ±0.22
DENIS-P 1026 − 0637	18.00 ±0.11	2.99 ±0.14	-	-
DENIS-P 2052 − 5512	17.52 ±0.13	2.70 ±0.15	0.74 ±0.23	3.44 ±0.26
DENIS-P 0142 − 4715	18.40 ±0.16	2.86 ±0.23	-	-
DENIS-P 0142 − 3952	17.80 ±0.16	2.72 ±0.18	-	-
DENIS-P 0020 − 4414	18.32 ±0.16	3.35 ±0.17	1.37 ±0.16	4.72 ±0.22
DENIS-P 0021 − 4244	16.83 ±0.05	3.20 ±0.05	1.38 ±0.06	4.58 ±0.08
DENIS-P 0151 − 6430	17.39 ±0.12	2.93 ±0.13	1.03 ±0.21	3.96 ±0.23
DENIS-P 0429 − 6649	17.32 ±0.13	2.89 ±0.14	1.33 ±0.17	4.22 ±0.21
DENIS-P 0430 − 8314	18.33 ±0.23	3.31 ±0.29	-	-
DENIS-P 0205 − 0653	16.83 ±0.06	2.89 ±0.07	1.27 ±0.11	4.16 ±0.12
DENIS-P 0205 − 1159	18.30 ±0.24	3.67 ±0.25	1.63 ±0.21	5.30 ±0.31
DENIS-P 0205 − 1637	18.64 ±0.22	3.04 ±0.27	-	-
DENIS-P 0449 − 0228	17.50 ±0.10	3.00 ±0.11	1.23 ±0.23	4.23 ±0.24
DENIS-P 0449 − 2225	18.70 ±0.20	2.84 ±0.26	-	-
DENIS-P 0247 − 1055	18.37 ±0.18	3.00 ±0.19	-	-
DENIS-P 0205 − 4313	17.95 ±0.15	2.78 ±0.21	-	-
DENIS-P 0205 − 3357	17.13 ±0.08	2.80 ±0.09	0.87 ±0.21	3.67 ±0.22
DENIS-P 2146 − 2153	18.40 ±0.27	2.98 ±0.29	-	-
DENIS-P 2143 − 8337	16.21 ±0.05	2.77 ±0.06	1.00 ±0.11	3.77 ±0.12
DENIS-P 2040 − 5700	18.14 ±0.17	2.90 ±0.20	-	-
DENIS-P 2040 − 3245	17.86 ±0.16	2.97 ±0.17	-	-
DENIS-P 0426 − 5735	18.45 ±0.20	3.17 ±0.22	-	-

3. And many more to come

In addition to the three BD candidates (including the confirmed BD DENIS-P J1228-1547) which are discussed above, the DENIS mini-survey has also provided a large sample of VLMS, and Table 2 lists the 35 dwarfs whose spectral type is later than M7V. Five of them have spectral types of M9V or later, doubling the size of the known inventory of such late objets.

From scaling of the DENIS mini-survey results (based on only 1% of the southern sky), it is clear that applying the same strategy to the full DENIS survey will eventually identify a few thousand M dwarfs later than M7V, a few hundred M dwarfs later than M9V, and perhaps 150 objects with the new L spectral type. Many more will only be detected in the more sensitive band, J, but from the DENIS data alone they will be very difficult to separate from the unavoidable cosmic ray hits and electronic glitches. With additional optical data, for instance from the Sloan Digital Sky Survey, they could be validated. This would then either yield some isolated equivalents of Gl 229B, whose colour would prevent a detection at I and K, or a significant upper limit on their local density. For 2-colour late dwarf detection (as opposed to identification), 2MASS has a significantly better sensitivity at J+H than DENIS has at I+J. Given the low discriminating power of [J-H] and [H-K], they however have to use additional data. Their initial plan is to use digitized Schmidt plates to separate late M dwarfs and BDs from earlier type stars, but combination with either DENIS data or the Sloan CCD survey will eventually have greater discriminating power.

Even for the smaller DENIS I+J catalog, the necessary follow-up will be a large work. High resolution optical spectroscopy will be needed for all objects to apply the Li test, and due to the optical faintness of these objects can probably only be obtained on 8m class telescopes. Parallaxes will also have to be obtained for a significant fraction of the sample, to, at last, obtain a reliable and unbiased low end of the local stellar luminosity function.

References

Allard F., Hauschildt P.H., Alexander D. R., Starrfield S. (1997) *ARAA*, **Vol. no. 35**, pp. 137-177

Becklin E.E, Zuckermann B. (1988) *Nature*, **Vol. no. 336**, pp. L656–659

Borsenberger J., (1996), Proc. 2nd Euroconfeence on Near-IR surveys (Tenerife), Kluwer ASSL Ser. **vol.210** pp. 181-186

Chabrier G., Baraffe I. (1997) *AA* **in press**

Delfosse X., (1997), Thèse de doctorat, Université de Grenoble 1

Delfosse X., Tinney C.G., Forveille T., Epchtein N., Bertin E., Borsenberger J., Copet E., de Batz B., Fouqué P., Kimeswenger S., Le Bertre T., Lacombe F., Rouan D., Tiphène D. (1997) *AAL*, **Vol. no. 327**, pp. L25-L28

Kirkpatrick D. (1998) This volume

Kulkarni S.R. (1997) *Science*, **Vol. no. 276**, pp. L1350–1354

Martin E.L., Basri G., Delfosse X., Forveille T. (1997) *AAL*, **Vol. no. 327**, pp. L29-L32

Nakajima T., Oppenheimer B.R., Kulkarni S.R., Golimowski D.A., Matthews K., Durrance S.T. (1995) *Nature*, **Vol. no. 378**, pp. 463–465

Rebolo R., Zapatero–Osorio M.R., Martin E.L. (1995) *Nature*, **Vol. no. 377**, pp.129–131

Rebolo R., Martin E.L., Basri G., Marcy G.W., Zapatero–Osorio M.R. (1996) *ApJ*,

   Vol. no. 327, pp. L53–L56
Reid N. (1994) *ASpSc.*, **Vol. no.217**, pp.57-62
Ruiz M.T., Leggett S.K., Allard F. (1997) *ApJL*, **Vol. no. 491**, pp. L107–L110
Thackrah A., Jones H., Hawkins M. (1997) *MNRAS*, **Vol. no. 284**, pp. 507–512
Stevenson D.J. (1991) *ARAA*, **Vol. no. 29**, pp. 163–193
Tinney C., Delfosse X., Forveille T. (1997) *ApJ*, **Vol. no. 490**, pp. L95–L98
Tinney C.(1995) *Mem. Soc. Ast. It.*, **Vol. no. 66**, pp. 611–618
Zuckermann B., Becklin E.E (1987) *ApJ*, **Vol.no. 319**, pp.L99–L102

SEARCHING FOR LOW-MASS STARS AND BROWN DWARFS WITH 2MASS

J. DAVY KIRKPATRICK
Infrared Processing and Analysis Center
California Institute of Technology
M/S 100-22, Pasadena, CA 91125, USA

MICHAEL F. SKRUTSKIE
Univ. of Massachusetts
Five College Astronomy Department, Dept. of Physics and Astronomy, Amherst, MA 01003, USA

JAMES LIEBERT, KEVIN LUHMAN AND MATT FISHER
Univ. of Arizona
Steward Observatory, Tucson, AZ 85721, USA

ROC CUTRI, CHARLES BEICHMAN AND CAROL LONSDALE
Infrared Processing and Analysis Center
California Institute of Technology
M/S 100-22, Pasadena, CA 91125, USA

NEILL REID
California Institute of Technology
M/S 105-24, Pasadena, CA 91125, USA

AND

DAVE MONET AND CONARD DAHN
U.S. Naval Observatory
P.O. Box 1149, Flagstaff, AZ 86002-1149, USA

1. Introduction

We have begun to look for low-mass stars and brown dwarfs using data from the 2MASS Prototype Cameras. Reported here are some very early, yet very encouraging, results which demonstrate the ease with which 2MASS will uncover missing, very cool members of the solar neighborhood.

N. Epchtein (ed.),
The Impact of Near-Infrared Sky Surveys on Galactic and Extragalactic Astronomy, 109-113.
© 1998 *Kluwer Academic Publishers.*

2. Search Method

Our search of the 2MASS databases thus far has been restricted to $b > 20°$ (to avoid confused regions near the plane) and to $K_s < 14.5$ (to insure brighter targets which could be followed up spectroscopically in the optical). Known brown dwarfs and brown dwarf candidates have colors as red as $J - K_s \approx 1.7$ for GD 165B and as blue as $J - K_s \approx -0.1$ for Gl 229B. Thus, $J - K_s$ color alone cannot be used as a selection criterion since most of the stars on the main sequence *also* fall in this range. However, a combination of colors from the optical and the near-infrared provide excellent criteria — known low-mass stars and brown dwarfs follow a monotoncially increasing sequence (e.g., in $R - K_s$) as temperatures decrease. Low-mass M dwarfs like LHS 2924 (M9 V) have $R - K_s \approx 7.5$ and the brown dwarf Gl 229B has $R - K_s \approx 10.8$.

Therefore, we have been pairing up the 2MASS database with digitizations of optical sky surveys in an effort to find 2MASS sources with no optical counterparts. For the data from the Prototype Cameras, we have used the APM digitization of the POSS-I plates, which have $R_{lim} \approx 20.0$. (For the actual survey, we will be using USNO digitizations of the POSS-II plates, which probe to $R_{lim} \approx 22.0$).

The pairing is done by position only, generally using a $3''$ matching radius. Thus, 2MASS targets with no matching sources in the optical fall into two categories: (1) proper motion objects which have moved more than $3''$ during the time interval between the optical and 2MASS surveys, and (2) infrared sources lacking *any* optical counterparts, even after possible motion matches were considered. Sources were retained if their $R - K_s$ colors or color limits exceeded 6.00, corresponding very roughly to a spectral type of M7 V.

3. Preliminary results

Using this search method, we have chosen a handful of low-mass star and brown dwarf candidates from the Protocam data. As of the date of this conference, we have obtained optical spectroscopy of sixteen sources. These are listed in Table 1 along with spectral types, K_s magnitudes, and estimated distances. Figure 1 shows the optical spectra of the objects obtained at the 5m Palomar and 4.5m MMT Observatories. A note on a few of the more interesting objects is given below.

3.1. 2MASP J1716586+451216AB

This object was first announced by Chester et al. (1992), where it was shown at higher resolution to be a close ($\sim 0.5''$), equal-magnitude binary. Our

TABLE 1. Protocam Sources with Follow-up Spectroscopy

Object Name[a]	Spectral Type	K_s mag	Estimated Distance (pc)
2MASP J1007435+113432	M4.5 V	12.77	102
2MASP J1244316+254720	M5.5 V	13.62	100
2MASP J1716586+451216AB	M6 V	13.14	83
2MASP J0152095+340037	M6.5 V	13.58	67
2MASP J0338549+223153	M6.5 V	14.34	83
2MASP J1246094+291603	M6.5 V	13.62	64
2MASP J1254369+253850	M7.5 V	13.24	48
2MASP J1524248+292535	M7.5 V	10.18	11
2MASP J1256227+283047	M7.5 V	13.53	55
2MASP J0339527+245728	M8 V	11.61	22
2MASP J0354012+231635	M8 V	11.79	24
2MASP J2234330+291850	M8: V	13.99[b]	37
2MASP J1242464+292619	M8: V	13.23	45
2MASP J1519431+260937	M8 V	14.33	73
2MASP J1520477+300210	M8.5 V	13.85	53
2MASP J0345432+254023	$\geq$M10 V	12.68	$\leq$21

[a] Source designations for 2MASS discoveries are given as "2MASx Jhhmmss[.]s±ddmmss." The "x" in the prefix will vary depending upon the catalog from which the object is taken; "P" is used for sources from the Prototype Camera data. The suffix conforms to IAU nomenclature convention and is the sexigesimal J2000-equinox RA and Dec, which for Protocam ("2MASP") sources is epoch 1993.5.
[b] This is the J-band magnitude.

joint spectrum of the pair is typed as M6 V. At the estimated distance of 83 pc, the physical separation on the sky is ~40 AU, meaning that this pair is, regrettably, too wide to provide an orbital determination and dynamical mass measures on a short timescale.

3.2. 2MASP J0338549+223153

As reported in Kirkpatrick, Beichman, & Skrutskie (1997), this object has a distance of ~83 pc assuming that it has an age comparable to an average field M dwarf (~1 Gyr). However, it lies in the direction of the Pleiades, and if it's a member of this cluster (age ~75 Myr), then it will be more luminous than a field M6.5 dwarf by ~0.8 mag at K_s. Its derived distance would then be ~125 pc, which is consistent with the distance to the Pleiades. Thus, this object may be a Pleiad lying near the cluster's stellar/substellar break, much like PPl 15 (Stauffer, Hamilton, & Probst 1994; Basri, Marcy, & Graham 1996).

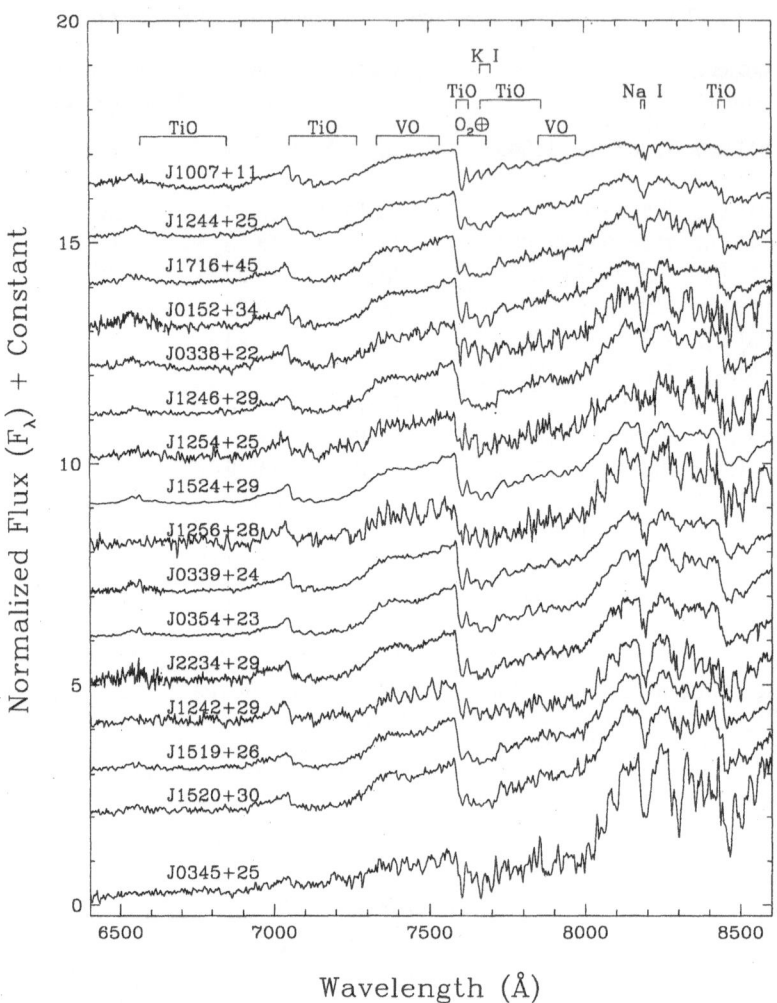

Figure 1.   Spectra from 6400 to 8600 Å of 2MASS Protocam sources. The sources are in the same order as that shown in Table 1, with the M4.5 dwarf at the top and the ≥M10 dwarf at the bottom. Note the increasingly red continuum at cooler and cooler temperatures along with the strengthening of the vanadium oxide (VO) bands. The 2MASS names have been shortened here for ease of plotting.

3.3. 2MASP J1524248+292535

Of the objects so far uncovered by 2MASS, this one — at an estimated distance of only 11 pc — is the closest new discovery to the Sun. Though it is easily detected on the POSS-I plates ($R \approx 17.5$) and has a large motion ($\mu \approx 0.61''/yr$), it was missed by Luyten during his proper motion survey of the northern sky (Fisher et al. 1997). This is the first of many such "missing" members of the solar neighborhood that 2MASS will uncover during the course of the survey.

3.4. 2MASP J0345432+254023

As of the date of this conference, this dwarf — of type $\geq$M10 — is the coolest object so far detected by 2MASS (Kirkpatrick, Beichman, & Skrutskie 1997). Though located toward the Pleiades, it is almost certainly a foreground object. The distance estimate, based on the trigonometric parallax for the slightly warmer ($\geq$M9) dwarf LP 944-20, places it at $\leq$21 pc.

4. Discussion

The results reported here are all based on Prototype Camera observations. As a first test, we have primarily chosen brighter objects for follow-up and thus have yet to probe into the deeper 2MASS data. Despite this, we have shown that 2MASS is already detecting extremely cool dwarfs, brown dwarf candidates toward the Pleiades, and missing low-mass members of the nearby stellar census.

The actual 2MASS survey is now underway. As we continue to build up data over larger areas, we will continue to scour them for late dwarf and brown dwarf candidates. Access to the Keck 10m telescope and low-resolution spectrograph will allow us to scrutinize even the faintest 2MASS sources. Follow-up on some of our early survey candidates will allow us to better understand our selection criteria, and our goal is to produce an all-sky catalog of "rare object" targets — including late-type M dwarfs, brown dwarf suspects, obscured carbon stars, IR-bright QSO candidates, etc.

Stay tuned for increasingly more exciting science from the Two-Micron All-Sky Survey.

References

Basri, G., Marcy, G. W., & Graham, J. R. 1996, ApJ, 458, 600

Chester, T., Beichman, C., Evans, T., Kopan, G., Schombert, J., Kleinmann, S., Lysaght, M., Skrutskie, M., Armus, L., Matthews, K., Neugebauer, G., Reid, N., Soifer, T., Tinney, C., & Hawley, S. L. 1992, BAAS, 181, 6809

Fisher, M., Liebert, J., Kirkpatrick, D., Cutri, R., Beichman, C., Reid, N., Djorgovski, G., & Monet, D. 1997, BAAS, in press

Kirkpatrick, J. D., Beichman, C. A., & Skrutskie, M. F. 1997, ApJ, 476, 311

Stauffer, J. R., Hamilton, D., & Probst, R. G. 1994, AJ, 108, 155

DENIS SURVEY OF AGB AND TIP-RGB STARS
IN THE LMC BAR WEST AND OPTICAL CENTER FIELDS

C. LOUP
Institut d'Astrophysique de Paris (IAP), CNRS

P.A. DUC, P. FOUQUÉ AND E. BERTIN
European Southern Observatory (ESO)

AND

N. EPCHTEIN
DESPA, Observatoire de Paris

Abstract. We have observed with the DENIS instrument two regions of $22' \times 23'$, located in the Bar of the LMC, and overlapping the Bar West and Optical Center fields studied by Blanco et al. (1980, BMB) in a prism-objective survey. Observations were performed simultaneously in the I, J, and K_S bands. There were 376 AGB stars identified in previous studies in these 2 fields. We typically find back 97% of them in the DENIS data. We show that the BMB's M and C stars follow two different sequences in the (I–J,J–K_S) colour-colour diagram. For M stars, the [I–J] colour is an indicator of the M subtype, increasing from 0.9 to 1.8 as one goes from M0 to M7 spectral types, while the [J–K_S] colour remains almost constant (from 1.0 to 1.35). Conversely, C stars are located on a "reddening" branch, with [J–K_S] ranging from 1.0 to about 2.2. M and C stars discovered by BMB have no or faint mass-loss. The (I–J,J–K_S) diagram allows to define a simple criterion to select stars of spectral types M and C. With this criterion, we find a total of 1177 new AGB and tip-RGB stars candidates, and have calculated the bolometric luminosities of all of them as well as previously known stars, assuming a distance modulus of 18.55. 90% of the new candidates are blue sources, not searched for in a complete way by BMB. Their luminosity distribution shows a very clear peak at $M_{bol} \simeq -3.5$, that we interpret as the tip of the Red Giant Branch. A second group is formed by stars for which the BMB survey should have been complete (i.e. relatively red C stars and M stars with spectral type later than M5.5). We find, however, 23% and 38% more such stars than they do in the Bar West and Optical Center, respectively.

N. Epchtein (ed.),
The Impact of Near-Infrared Sky Surveys on Galactic and Extragalactic Astronomy, 115-127.
© 1998 *Kluwer Academic Publishers.*

Finally, we also discover in this study 42 red sources, with $2.2 < [J-K_S]$ < 4.2. These stars experience mass-loss at significant rates. Note that there might be redder obscured AGB stars in the fields, but that the DENIS limiting magnitudes do not allow to detect sources with $[J-K_S]$ larger than about 4. Only 4 of them have an IRAS counterpart. It is interesting to note that we discover 42 mass-loosing AGB stars in 0.14 square degrees, while only about 50 have been discovered in the whole LMC using the IRAS data. The luminosity distribution of these red sources is very similar to the one of the BMB's C stars, ranging from $M_{bol} \simeq -3.5$ to -6.

1. Introduction

Searches for AGB stars and red supergiants in the Magellanic Clouds have started in the sixties, through optical/I photographic and prism-objective surveys. One of the most fundamental one is the work of Blanco et al. (hereafter BMB) in 1980. Compared to previous works which were (almost) spatially complete, they observed only 3 small regions in the Large Magellanic Cloud (LMC), and 2 in the Small Magellanic Cloud (SMC), of 23' in diameter. However, their I limiting magnitude was 17 instead of 14 in previous works, so that they typically found 50 times more AGB stars than previously known. Their survey was later extended to 52 regions in the LMC (Blanco & McCarthy 1990, BM). The main result of their surveys was to derive the true luminosity function of LMC carbon stars (see e.g. Richer 1981, and Costa & Frogel 1996), while only the brightest tail was known before. This brought a very serious problem to the theoreticians as theories predicted much too high a luminosity for C stars. The problem is getting solved by taking into account physical phenomena like high mass-loss rates and "overshotting" (see e.g. Bertelli et al. 1985).

This type of surveys, as well as surveys devoted to the search for Long-Period Variables (LPVs, see e.g. Hughes 1989), does not allow to find AGB stars with high mass-loss rates as they become invisible in the optical/I. Actually, these surveys picked up only AGB stars without, or with faint, mass-loss. One of the first assumption for the lack of high luminosity C stars predicted by theories was that they could all loose mass at a high rate, and so have been missed by BMB. The JHK survey of Frogel & Richer (1983) however failed to find such very luminous obscured sources. Later, searches for "dust obscured" AGB stars and red supergiants became more intensive with the use of the IRAS data. At the moment, about 50 obscured AGB stars have been discovered thanks to IRAS data (see e.g. Reid et al. 1990, Wood et al. 1992, Zijlstra et al. 1996). It shows that some very luminous obscured AGB stars exist, but they are rare. The main limitation of IRAS observations in the Magellanic Clouds is that source fluxes are very close to the detection limit. As a consequence, only the most luminous and reddest sources have been found, and even for such sources IRAS is not complete (Loup et al. 1997). Between optical/I and IRAS surveys, we still end up with a very incomplete view of the AGB population in the Magellanic Clouds. First, the whole optically thin population, including M and C stars, is actually well known in only 1 BMB field, the Bar West field (Frogel & Blanco 1990). Second, all the population with intermediate mass-loss rates is missing as it has not been seen neither in optical/I surveys, nor by IRAS.

The goal of the present work was to estimate to which extent the IJK_S DENIS survey could provide us with a much more complete knowledge of the LMC AGB population. It has been concentrated on two fields observed by BMB : the Bar West (BW) and Optical Center (OC) fields. In section 2 we briefly present the DENIS observations. In section 3, we present the cross-identification between BMB and DENIS sources, and the location of these stars in the $(I–J, J–K_S)$ colour-colour diagram as a function of spectral type. In section 4, we present new DENIS AGB star candidates not found by BMB, in particular those with significant mass-loss, and the resulting luminosity distribution. Finally, in section 5, we give the main conclusions.

2. Observations

Observations have been performed with the 3 cameras of the DENIS instrument mounted at the ESO 1m telescope (la Silla), simultaneously in the I, J, and K_S bands, in December 1995. The field of view is $12'$, the pixel size $1''$ in I and $3''$ in J and K_S. Integration times were 9s in J and K_S, and only 1s in I (it is 9s too in the normal survey mode). Our goal was to observe two fields previously surveyed by BMB in the Bar of the LMC : the Bar West and the Optical Center fields. For each field, we performed a mosaic of 6 images, 2 images Est-West and 3 North-South, with an overlap of $2'$, so that the total mapped area is 22x32'. The mosaic has been repeated 3 times within 15 minutes the same night on both fields. The DENIS fields do not overlap perfectly with the BMB fields : we miss a few of their sources in the North-East and North-West directions. Magnitudes of point sources have been determined with the "Sextractor" software (Bertin & Arnouts 1996). Confusion limit is not reached in Bar West (except in globular clusters), nor in K_S in the Optical Center; some parts of the field are rather crowded in I and J in the Optical Center.

In this study, we only kept point sources which are, at least, detected in the K_S band (so sources detected only in I and J are excluded), not blended, confirmed in the 3 repeated mosaics, and with an uncertainty on magnitudes better than 0.15 mag. This last requirement, and the short integration time in the I band, limits the sensitivity of this study to magnitudes 15.5 in I, 15 in J, and 12.8 in K_S.

3. Previously known AGB stars

In the following we limit the discussion to stars in the fields commonly observed by BMB and DENIS, i.e. to two fields of 22x23'. In the Bar West (BW) field, BMB found 40 M stars with spectral types later than M5.5, and 65 C stars. Later, Frogel & Blanco (1990, FB) found 96 M stars with spectral types earlier than M5.5, 1 additional C star, and finally 50 other M stars in a third, more complete, search performed in 30% of the BMB-BW field. The BW field also contains 2 M stars (Rebeirot et al. 1983), and 4 Mira and 2 SRa variables (Hughes 1989, spectral type unknown) which were not found by BMB. In the Optical Center (OC) field, BMB found 32 M stars with spectral type later than M5.5, and 73 C stars. There are 5 additional M stars (Rebeirot et al. 1983, Westerlund et al. 1981, 1975), and 3 Mira and 6 SRa variables of unknown spectral type (Hughes 1989), not found by BMB. In total 186 and 37 M stars, 65 and 73 C stars, and 6 and 9 LPVs of unknown spectral type, were known in the BW and OC fields respectively.

Cross-identifications of these stars with DENIS sources were done on the basis of coordinates (when available) and finding-charts (especially in crowded parts). Excluding the third more complete search for M stars of FB, we find back 99% and 96% of the previously known sources in the BW and OC fields respectively. The few sources not recovered are located in crowded parts or coincide with bad pixels. For the third search of FB, the recovering rate is only 64%. The M stars of FB which are not recovered are fainter and not detected in the K_S band; they are however detected in the I and J bands.

DENIS allows for the first time to study the location of AGB stars in the $(I-J, J-K_S)$ colour–colour diagram as observations are performed simultaneously in the 3 bands. Such a diagram made with previous non-simultaneaous observations actually suffers from a large spread (Loup & Groenewegen 1994), due to the variability of AGB stars, and due to the differences between the photometric systems. The $(I-J, J-K_S)$ DENIS diagram is presented in Figure 1, for both BW and OC fields together as no differences in the source locations were found. Fig.1a shows the location of all the DENIS sources selected as defined in section 2. One can see already that the stars are located on two branches. The main branch starts at about (0,0), goes towards increasing values of both colours and ends up with a vertical branch, $[J-K_S]$ beeing almost constant while $[I-J]$ increases. The

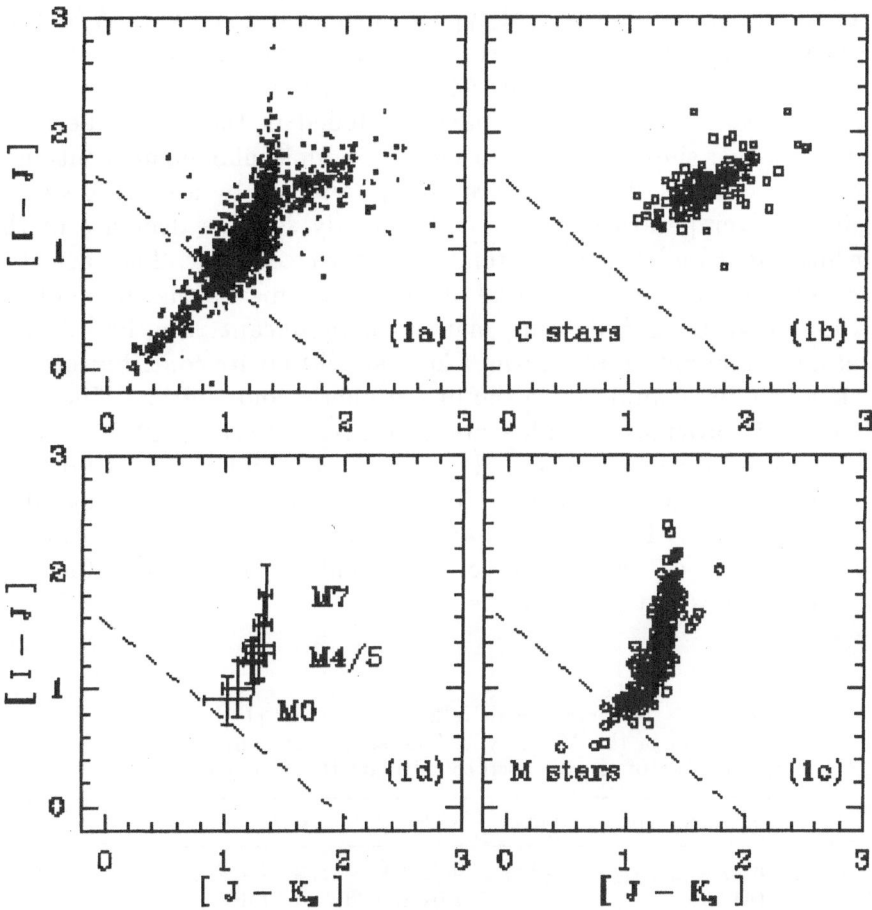

Figure 1.   The DENIS (I–J,J–K$_S$) colour–colour diagram. Fig.1a shows the location of all the DENIS sources in both BW and OC fields, with an error on the magnitude smaller than 0.15 in the 3 bands. Fig.1b and 1c show the location of the BMB's C and M stars, respectively. Fig.1d presents the mean colours of the BMB's M stars, derived for each M subtype (see also Table 1). The dashed line in the 4 plots indicates the criterion used to select AGB and tip-RGB stars from the DENIS data (section 4).

second branch, located in the reddest part of the diagram, seems to follow a redening sequence. In Fig.1b and 1c, we show the location of the BMB C and M stars, respectively. Clearly, the red branch is formed by C stars, while M stars form the vertical branch. However, the two branches have a limited overlap around I-J=1.5 and J-K=1.4. It has been shown (Groe-

newegen private communication, Loup & Groenewegen 1994) that the main
branch of the diagram is a spectral type sequence, starting with A/B type
until M type stars. The vertical part of the main branch is also a spectral
type sequence, from M0 to M8/10. As one goes towards later M spectral
types, the TiO and VO molecular bands included in the I filter becomes
deeper. On another side, there are no such huge molecular bands contained
in the J and K_S filters. J and K_S are mainly affected by a small decrease
of the effective temperature. So what we mainly see is a decrease of the
I flux, while the J and K_S fluxes remains almost constant. The fact that
[J–K_S] ranges in a so narrow interval of values also means that none of the
M stars found by BMB is loosing mass at a significant rate, i.e., if they
loose mass, the amount of dust formed is so small that its contribution to J
and K_S is negligible compared to the photospheric contribution. The work
of BMB and FB provides enough stars from spectral types M0 to M7.5 to
allow us to make statistics on the [I–J] and [J–K_S] colours as a function
of the M subtype. The mean colours for each M subtype are presented in
fig.1d, and in Table 1. It confirms that [I–J] increases as one goes towards
later M spectral types. In the future, this could be helpful to determine,
statistically, the M spectral type index.

TABLE 1. Observed mean values of [I–J] and [J–K_S] as
a function of the M subtype. The second column gives
the number of sources used to calculate the averages.

Spectral type	Number	$< I - J >$	$< J - K_S >$
M0/0.5	14	0.91 ± 0.21	1.02 ± 0.20
M1/1.5	33	1.01 ± 0.23	1.11 ± 0.13
M2/2.5	20	1.23 ± 0.19	1.22 ± 0.11
M3/3.5	17	1.24 ± 0.19	1.25 ± 0.09
M4/4.5	13	1.31 ± 0.24	1.29 ± 0.12
M5/5.5	27	1.37 ± 0.27	1.29 ± 0.13
M6/6.5	34	1.55 ± 0.27	1.32 ± 0.07
M7/7.5	22	1.81 ± 0.25	1.34 ± 0.05

C and M stars are distributed almost in opposite ways, in the sense
that C stars exhibit a larger range of colours in [J–K_S] than in [I–J]. At
first glance (see also Loup & Groenewegen 1994), the C stars seems to
follow a reddening sequence which could be attributed to circumstellar
dust extinction. However, if it would be so, it would mean that many C
stars discovered by BMB loose mass at a significant rate, while M stars
do not. We have not seen any systematic bias in their observations which

could explain that. It is not in agreement with observations in our Galaxy where M stars are observed with all the range of mass-loss rates. Finally, M and OH/IR stars loosing mass at significant and high rates have been discovered in the LMC (Wood et al. 1992, Zijlstra et al. 1996). It is thus probably not the right hypothesis to assume that the red C star branch is due to mass-loss. We now take as an assumption that C stars seen by BMB do not loose more mass than the BMB's M stars. $[J-K_S]$ ranges from 1.0 to 2.2 (plus 3 objects between 2.2 and 2.5). The first effect one can think of is the influence of the effective temperature. We have calculated, using the real DENIS passbands (Fouqué et al. 1997), that blackbodies with temperatures of 2500 and 2000 K would have $[J-K_S] = 1.4$ and 1.9, respectively. So, assuming that C stars may have effective temperatures as low as 2000 K, their $[J-K_S]$ values may be as large as 2. However, as shown by Cohen et al. (1981), the effect of blanketing by molecules like C_2, CN, and CO, between 1 and 3 μm, has a large influence on the resulting $[J-K]$ colour. They demonstrate that, for an effective temperature of 3000 K, one can easily reach $[J-K]$ of the order of 2. Hence, we conclude that both C and M stars discovered by BMB have no or faint mass-loss, and that the red branch formed by C stars in the $(I-J, J-K_S)$ diagram is due to the combined influence of the effective temperature and carbon molecule blanketing. The few C stars with $[J-K_S]$ larger than 2.2 might start to be also affected by circumstellar dust.

We finally would like to end this section with a remark. Figure 1 could give the impression to the reader that the "C" branch contains only C stars. Let say for instance that all the stars with $[J-K_S] > 1.6$ would be C stars. This would be equivalent to similar conclusions derived in the past from (J–H,H–K) diagrams. We think, however, that using the diagram in that way might be erroneous because it neglects the effect of the mass-loss. As soon as an M star starts to loose mass at a significant rate, its location in the diagram will be shifted to the red part, following a reddening line which is almost parallel to the C star branch (at least for not high mass-loss rates). One example of such a source with $[J-K_S] > 1.6$ is displayed in Figure 1c. This star is a known supergiant, WOH-SG 264 (Westerlund et al. 1981), with an IRAS counterpart, showing that the star does loose mass. In Figure 2, one can see also that we find quite a few new AGB stars candidates with $1.6 < [J-K_S] < 2.2$, not found by BMB (see section 4). They can be *a priori* C stars, or M stars with some mass-loss. Spectroscopic follow-up is of course required for these sources. The only conclusion that we may draw out is that stars located on the vertical branch and above the C star branch are very likely to be of late M–type (Fig.1a).

4. New DENIS AGB star candidates

Figure 1 allows to define a simple criterion to select new AGB star candidates from the DENIS data. We have taken all the sources located above the dashed line, and not identified in previous studies (see section 3, mostly from BMB and FB). We put the line a little below the mean colours of M0 spectral types. It means that we have in principle selected most M stars, and all the C stars. As seen in Fig.1c, we have probably missed a few M0/1 stars. We have also probably selected a few late K type stars. In addition to AGB stars, our selection has also picked up a number of RGB stars close to the RGB tip. In total, following this selection criterion, we find 369 and 808 new AGB or tip–RGB star candidates in the BW and OC fields, respectively.

These DENIS candidates can be divided in three groups. In the first group we put all the "blue" sources. According to figure 1, they might have spectral types from late K to M4/5, or they could be blue C stars with $[J-K_S] < 1.4$. About 90% of the new candidates belong to this group. In the second group we put the DENIS sources located on the C star branch defined by the BMB C stars with $1.4 < [J-K_S] < 2.2$, as well as stars located on the last part of the vertical branch formed by their M stars, i.e. sources expected to have spectral types later than M5.5 ($[I-J] > 1.5$). According to BMB, their survey should be complete for such objects. We find, however, 26 and 50 new DENIS candidates belonging to this second group in the BW and OC fields, respectively. Comparing with the BMB stars having the same properties, they missed 23% and 38% of the stars in the BW and OC fields.

Finally, we have put in the third group all the DENIS candidates with $[J-K_S] > 2.2$, i.e. AGB stars supposed to experience a significant mass-loss, that hampers optical/I surveys detection (except a few, see Figure 1). There are 10 in the BW field, and 32 in the OC field. Most of them have not been detected in the I band, because of circumstellar dust extinction, but also because our integration time in I was only 1s. The reddest $[J-K_S]$ colour is about 4.2. It does not mean that redder sources do not exist in the field. The limiting magnitudes are such that redder sources would not be detected in the J band. Very obscured AGB stars with $[J-K]$ as large as 6 have been found among the IRAS sources (Reid 1991, Zijlstra et al. 1996).

We have calculated the bolometric magnitudes of all the DENIS sources, including previously known BMB stars and new candidates, detected at least in the J and K_S bands. We assume a distance modulus of 18.55. Bolometric magnitudes have been calculated by integrating over the IJK_S fluxes with spline interpolation. To take into account the energy radiated at

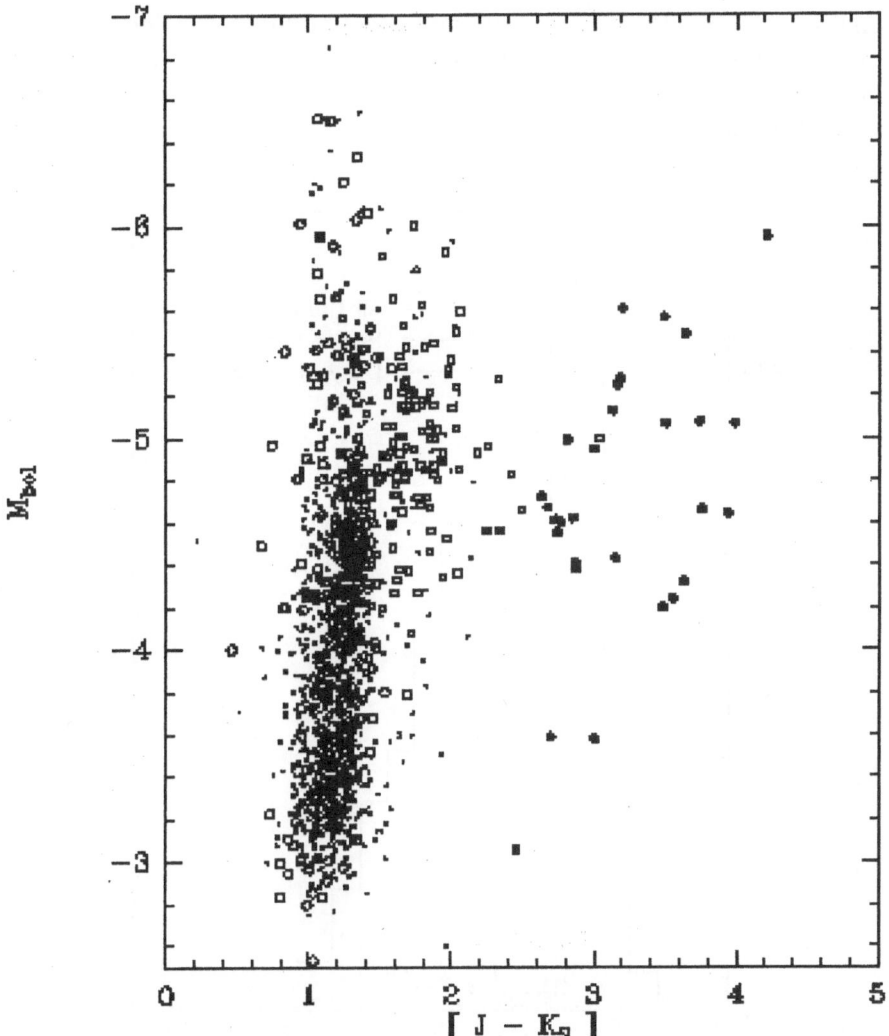

Figure 2.   Bolometric magnitudes as a function of the [J–K$_S$] colour. Open circles are previously known M stars, open squares previously known C stars, and open triangles previously known LPVs of unknown spectral type. Filled circles are the new AGB and tip-RGB DENIS candidates. Bolometric magnitudes have been calculated as explained in section 4, with a distance modulus of 18.55.

shorter wavelenghts than I and longer than K$_S$, a blackbody was fitted and integrated from 0.1 to 0.75 and from 2.2 to 60 μm. For sources with [J-K$_S$] > 2.7, we have however used the bolometric correction of Groenewegen (1997) to better take into account the energy radiated by the circumstellar dust. Groenewegen shows that the correction is negligible for smaller values of [J–K_S]. Finally, for the 4 IRAS sources, the flux was integrated by splines until 7.5 μm, and the IRAS contribution was calculated following the method of Loup et al. (1993).

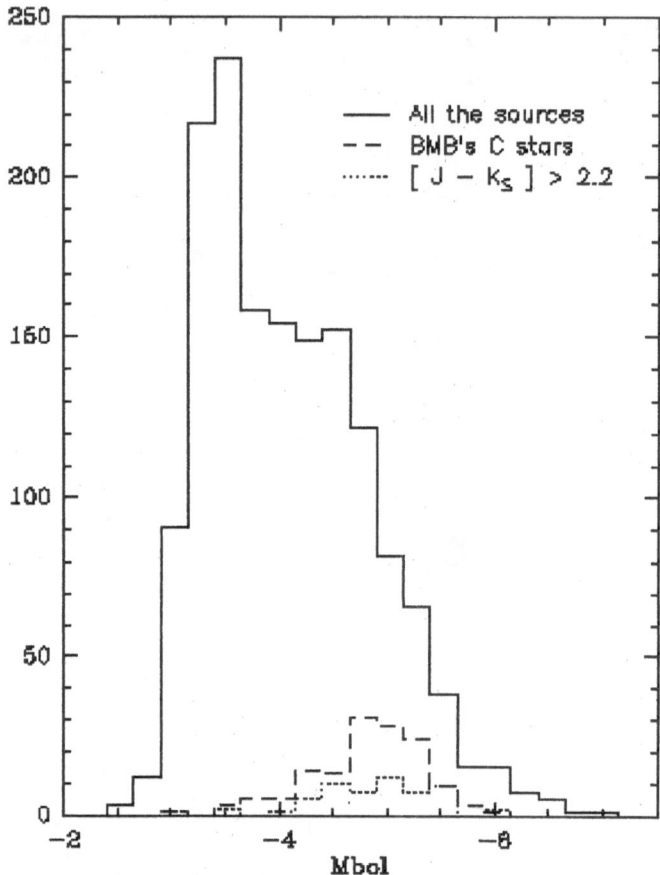

Figure 3. Luminosity distribution of the DENIS sources. Bolometric magnitudes have been calculated as explained in section 4, with a distance modulus of 18.55. The peak at $M_{bol} \simeq -3.5$ corresponds to the tip of the RGB sequence. The distribution of the new DENIS sources with significant mass-loss rates is very similar to the one of the optical BMB's C stars.

The results are shown in Figures 2 and 3. Figure 2 displays the (J–K_S,M_{bol}) colour-magnitude diagram. It can be compared to the diagram shown by Costa & Frogel (1996) for a sample of more than 800 C stars (mostly from BMB and BM). The distributions of the optical C stars in our diagram and Costa & Frogel's one are slightly different. In their Fig. 2, the average luminosity increases from [J–K] = 1.0 to 1.7, and decreases for larger values of [J–K], up to about 2.7. We find that the luminosity of most optical C stars is essentially independant of [J–K_S]. In addition, there are a few C stars with faint luminosities (M_{bol} < −4) and 1.0 < [J–K_S] < 1.7, which are not found by Costa & Frogel, although they do include these C stars in their sample. We think that the difference comes from their calculation of the bolometric fluxes derived from a bolometric correction based on [J–K] and a relatively small sample of stars. The main new feature in our colour-magnitude diagram is of course the presence of stars with large values of [J–K_S]. Figure 2 shows that their range of luminosities is similar to the one of the BMB's C stars. There are 3 stars with [J–K_S] > 2.2 and faint luminosities (M_{bol} ∼ −3.5). Their nature is still unclear; we do not exclude the possibility that they are RGB stars experiencing a phase of mass-loss (see below, and Fig.3).

In Figure 3, we present the luminosity distribution of all the DENIS sources, the BMB's C stars, and our new candidates with [J–K_S] > 2.2. In the distribution of all the sources, there is a very clear peak at about M_{bol} = −3.5. The most likely interpretation is that we find the tip of the Red Giant Branch, so that a non negligible percentage of our new "blue" DENIS candidates (first group) are tip-RGB stars rather than AGB stars on the E-AGB branch. The drop of the distribution at lower luminosities only reflects the limits of our study. It will be interesting in the future to also keep sources detected only in I and J, as one should thus find faint AGB stars at the very begining of the E-AGB branch, as well as fainter RGB stars. The brightest tail of the distribution is mainly formed by M0 to M2 stars (see also FB). These stars probably belong to luminosity classes I or II. The luminosity distribution of the BMB's C stars is in good agreement with previous studies (see e.g. Cohen et al. 1981). More interestingly, we show that the distribution of the new DENIS candidates experiencing large mass-loss ([J–K_S] > 2.2) have essentially the same distribution than the BMB's C stars, and that none of them is more luminous than M_{bol} ≃ −6, including the 4 IRAS sources. This bring us back to the closing remark of section 3. Does it means that all these red sources are C stars ? If yes, it would mean that there are very few M stars loosing mass at significant rates in the LMC, and that most M stars in a certain range of initial masses would become C stars before to reach the stage of mass-loss. If not, it means that M and C stars experiencing mass-loss have very similar luminosity distributions. Spectroscopy of the red stars discovered in this study must be definitely performed.

5. Conclusions

We show that DENIS is the most powerful tool we ever had to study the AGB and RGB populations in the Magellanic Clouds. Including sources detected only in I and J, or only in J and K_S, DENIS allows to discover most of the optically thin AGB population, as well as a large fraction of the RGB population. We show that the tip of the RGB branch occurs at $M_{bol} \simeq -3.5$, assuming a distance modulus of 18.55. On the side of obscured AGB stars, DENIS allows to find them up to $[J-K_S]$ of the order of 4.0. More optically thick AGB stars will not be detected and should be searched for using other instruments, in particular the ISO satellite.

Optical C and M stars previously found by Blanco et al. (1980, BMB) and Frogel & Blanco (1990) in the so-called Bar West and Optical Center fields in the Bar of the LMC, appear well separated in two branches in the $(I-J, J-K_S)$ DENIS colour-colour diagram. For M stars, the $[I-J]$ colour is an indicator of the M subtype. This diagram also allows to define a simple criterion to select M and C stars. We find a total of 1177 new AGB and tip-RGB star candidates in both BMB's fields (excluding sources not detected in the K_S band), to be compared with 376 stars identified in previous studies. Though BMB reasonably thought to be complete in their survey of relatively red C stars and M stars with spectral types later than M5.5, we find 23% and 38% more such sources in the Bar West and Optical Center fields, respectively. The main new feature of this study is the discovery of 42 sources with $[J-K_S]$ ranging between 2.2 and 4.2, experiencing mass-loss at non-negligible large rates. Their luminosity distribution is very similar to the one of the BMB's C stars, ranging from $M_{bol} \simeq -3.5$ to -6. Only 4 of them have an IRAS counterpart, the brightest ones as expected. Their chemical type is still unknown, and spectroscopic follow-up should be performed.

References

Bertin E., Arnouts S., 1996, A&AS 117, 393
Blanco V.M., McCarthy M.F., 1990, AJ 100, 674 (BM)
Blanco V.M., McCarthy M.F., Blanco B.M., 1980, ApJ 242, 938 (BMB)
Cohen J.G., Frogel J.A., Persson S.E., Elias J.H., 1981, ApJ 249, 481
Costa E., Frogel J.A., 1996, AJ 112, 2607
Epchtein N., et al., 1997, *The Messenger* 87, 27
Fouqué P. et al., 1997, in preparation
Frogel J.A., Blanco V.M., 1990, ApJ 365, 168 (FB)
Frogel J.A., Richer H.B., 1983, ApJ 275, 84
Groenewegen M.A.T., 1994, A&A 288, 782
Hughes S.M.G., 1989, AJ 97, 1634
Hughes S.M.G., Wood P.R., 1990, AJ 99, 784
Loup C., Forveille T., Omont A., Paul J.F., 1990, A&AS 99, 291
Loup C., Groenewegen M.A.T., 1994, Ap&Sp.Sci. 217, 131

Loup C., Zijlstra A.A., Waters L.B.F.M., Groenewegen M.A.T, 1997, A&AS 125, 419

Rebeirot E., Martin N., Mianes P., Prévot L., Robin A., Rousseau J., Peyrin Y., 1983, A&ASS 51, 277

Reid I.N., 1991, ApJ 382, 143

Reid I.N., Glass I.S., Catchpole R.M., 1988, MNRAS 232, 53

Richer H.B., 1981, ApJ 243, 744

Westerlund B.E., Olander N., Hedin B., 1981, A&AS 43, 267 (WOH)

Westerlund B.E, Olander N., Richer A.B., Crabtree D.R., 1978, A&ASS 31, 61

Wood P.R., Bessel M.S., Paltoglou G., 1985, ApJ 290, 477

Wood P.R., Whiteoack J.B., Hughes S.M.G., Bessell M.S., Gardner F., Hyland A.R., 1992, ApJ 397, 552

Zijlstra A.A., Loup C., Waters L.B.F.M., Whitelock P.A., Guglielmo F., 1996, MNRAS 279, 32

DENIS AND HIPPARCOS:
LUMINOSITY AND KINEMATIC CALIBRATIONS

R. ALVAREZ AND M-O. MENNESSIER
GRAAL (ESA 5024/CNRS), Université Montpellier II,
F-34095 Montpellier Cedex 05, France

Abstract. The homogeneous set of apparent magnitudes in the three photometric bands that the DeNIS survey will provide may be fruitfully combined with the astrometric and kinematical data of the Hipparcos satellite in order to apply a maximum-likelihood method and derive absolute magnitudes, kinematics and spatial distribution for extensive samples of stars.

1. Introduction

The DeNIS survey and the Hipparcos mission both provide us a considerable amount of data. Sophisticated statistical treatments are necessary to fully exploit all this information and thus improve our knowledge on Galactic structure, populations classification, HR diagram calibration, etc. One of this possible treatment is the LM method (Luri, Mennessier et al. 1996) specifically developed to exploit the Hipparcos data to its full extent for luminosity calibrations and distance determinations. The DeNIS survey is of prime interest for the application of the LM method as it provides homogeneous photometry for a large sample of stars with available astrometric and kinematical data. Furthermore, the observational censorship used to define a sample is known (Hipparcos selection function + DeNIS selection function): this permits to derive almost unbiased estimates of absolute magnitudes and distances (Luri & Arenou 1997).

2. The LM method

The reader is referred to Luri, Mennessier et al. (1996) for a thorough description of the LM method. We outline here its most important features. This method, based on the maximum-likelihood principle, allows us to si-

N. Epchtein (ed.),
The Impact of Near-Infrared Sky Surveys on Galactic and Extragalactic Astronomy, 129-133.
© 1998 *Kluwer Academic Publishers.*

multaneously calibrate the luminosity and determine the mean kinematic characteristics and spatial distribution of a given sample. This sample is specifically modeled with appropriate distribution functions corresponding to the absolute magnitudes, kinematics and spatial distributions. Sampling effects, the galactic differential rotation and observational errors are rigorously taken into account by including appropriate functions in the density law describing the sample. The method is able to use inhomogeneous samples, i.e. samples composed of a mixture of groups of stars with different luminosities, kinematics or spatial distributions. In this case the method identifies and separates the groups. Moreover, the LM method assigns each star to a group and estimates its most probable distance.

3. Hipparcos stars with available DeNIS photometry

As any statistical method, the need for sufficiently large sample is essential. Relying on V magnitudes and V−I colours taken from the Hipparcos Catalogue (ESA 1997), we have counted how many stars among the over 120 000 Hipparcos Catalogue entries are fainter than 9.5 mag in I (DeNIS limit of saturation). The number of such stars was counted per spectral type and luminosity class. More than half the Hipparcos stars have not been attributed a luminosity class. Nevertheless, for each spectral type, we randomly dispatch the unclassified stars among the luminosity classes according to their relative weights. Finally, as an approximation of the southern-sky coverage of DeNIS, the number in each class was divided by two. The results are shown in Table 1. The B to M giants and B to G dwarfs appear to be the samples which contain enough stars ($\geq$ 100) to apply the LM method. If TYCHO data were used, these numbers should be multiplicated by a factor $\approx$ 8. The larger available samples would then compensate for the less accurate proper motion measurements.

4. Example of application of the LM method

We have considered a sample of 103 oxygen-rich Miras and 129 Semi-Regular variables (Alvarez et al. 1997) with available Hipparcos trigonometric parallaxes and proper motions (ESA 1997); radial velocities from the Hipparcos Input Catalogue (Turon et al. 1992); and apparent K magnitudes from Catchpole et al. (1979), Fouqué et al. (1992), Whitelock et al. (1994), Kerschbaum & Hron (1994) and Kerschbaum (1995). In order to describe the sample, the following distribution functions have been adopted:

1. **Distribution of absolute K magnitudes:** a gaussian law with mean M_0 and standard deviation σ_M

TABLE 1. Estimates of the number of Hipparcos stars
with available DeNIS photometry per spectral type and
luminosity class

Spectral Type	Dwarfs & Subdwarfs	Giants & Subgiants	Supergiants & Bright giants
O	9	2	4
B	211	140	56
A	353	127	20
F	919	144	28
G	360	278	21
K	94	386	9
M	43	126	7
R N S C	–	7	–

2. **Velocity distribution:** a Schwarzschild ellipsoid with means (U_0, V_0, W_0) and dispersions $(\sigma_U, \sigma_V, \sigma_W)$
3. **Spatial distribution:** an exponential disc with scale height Z_0

Three distinct classes of stars with different kinematics and scale heights have been identified. Table 2 gives the estimates of the model parameters in the columns marked θ and the corresponding errors are given in the columns marked σ.
Two significant populations are well separated. Group 1, which is the main one (about 81 % of the sample), has the kinematics of late disk stars. The scale height is characteristic of the old disk population. This group can be interpreted as the standard disk population which has an exponential scale height of $\sim$ 300 pc (Jura & Kleinmann 1992).
Group 3 (about 17 % of the sample) has a larger velocity ellipsoid. The scale height is much more important. The large velocity ellipsoid and the important scale height characterize a population older than the group 1. The stars of this group might belong to the extended/thick (E/T) disk or they might be halo stars.
Group 2 is a very small group with very low velocity dispersion. It might be formed by a sub-population of younger stars. The small number of stars prevents further interpretation.

The individual absolute magnitudes of the Miras are also obtained with the LM method. They are plotted against the periods (Fig. 1). The three

TABLE 2. Model parameters using m_K (103 Miras and 129 Semi-Regulars)

		group 1		group 2		group 3	
		θ	σ	θ	σ	θ	σ
M_0	(mag)	-6.3	0.7	-6.1	1.6	-6.7	0.9
σ_M	(mag)	1.0	0.4	0.4	0.4	0.8	0.5
U_0	(km.s^{-1})	-11	6	-53	17	-33	40
σ_U	(km.s^{-1})	37	8	1	7	93	34
V_0	(km.s^{-1})	-23	6	-57	74	-93	53
σ_V	(km.s^{-1})	22	4	15	30	75	24
W_0	(km.s^{-1})	-12	4	-33	10	-2	33
σ_W	(km.s^{-1})	20	5	3	4	58	31
Z_0	(pc)	260	40	370	180	820	240
%		81	7	2	1	17	7

groups are distinguished by different symbols. For the two most significant groups (1 and 3), least-square linear fits are obtained: two parallel period–luminosity relations are found in K, one for each population.

Figure 1 also shows the M_K—P relation that Van Leeuwen et al. (1997) have calibrated for Galactic oxygen–rich Miras by using Hipparcos parallaxes and adopting a priori the slope of the Large Magellanic Cloud (LMC) relation. Its slope is in very good agreement with ours. This is a very remarkable result: we find that the slopes of the Galactic period–luminosity relations in K are the same as the LMC one.

It has been discussed for a long time as to whether metallicity effects in Miras might generate different period–luminosity relations. The results of the present work tend to demonstrate that Galactic Miras follow different period–luminosity relations in K, according to the two distinct populations that we have separated: the slopes are the same and only the zero points differ by about 0.5 mag. The shift between the period–luminosity relations is interpreted as the consequence of the effects of metallicity abundance on the luminosity.

5. Conclusion

Hipparcos has provided a large number of trigonometric parallaxes and proper motions. DeNIS and 2MASS will provide precise and homogeneous photometric data allowing to define specific classes of stars, constituting homogeneous samples to which the LM method may be applied. This will

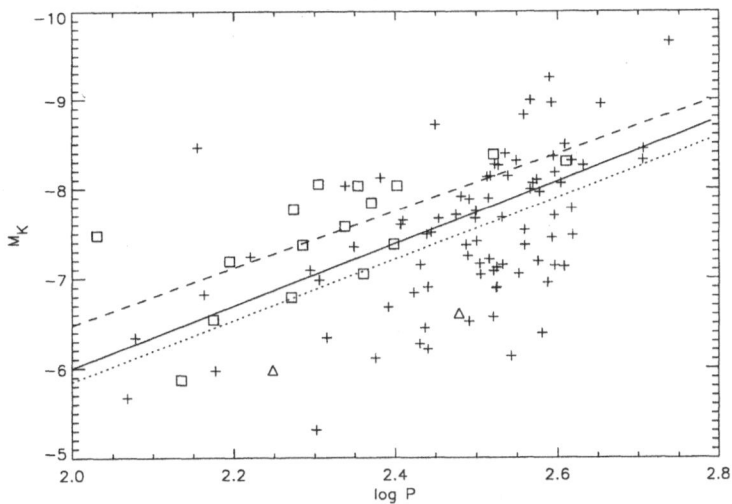

Figure 1.   Period–luminosity relations in K band. Crosses represent Miras belonging to group 1, triangles to group 2 and squares to group 3. Dotted line and dashed line are the period–luminosity fit relations for group 1 and 3 respectively. Solid line is the period–luminosity relation determined by Van Leeuwen et al. (1997)

enable us to calibrate the luminosities of large number of B to M giants and B to G dwarfs, as was done for the Long Period Variables.

References

Alvarez R., Mennessier M-O., Barthès D., Luri X., Mattei J.A., 1997, A&A in press

Catchpole R., Robertson B., Lloyd Evans T., Feast M., Glass I., Carter B., 1979, SAAO Circ 1, 61

ESA, 1997, The HIPPARCOS Catalogue, ESA SP-1200

Fouqué P., Le Bertre T., Epchtein N., Guglielmo F., Kerschbaum F., 1992, A&AS 93, 151

Jura M., Kleinmann S., 1992, ApJS 79, 105

Kerschbaum F., 1995, A&AS 113, 441

Kerschbaum F., Hron J., 1994, A&AS 106, 397

Luri X., Arenou F., 1997. *Utilisation of Hipparcos Data for distance determinations: error, bias and estimation.* In: Perryman M.A.C. (ed.) The Hipparcos and Tycho Catalogues, ESA-SP 402, in press

Luri X., Mennessier M-O., Torra J., Figueras F., 1996, A&AS 117, 405

Turon C., Crézé M., Egret D. et al., 1992, The HIPPARCOS Input Catalogue, ESA SP-1136

Van Leeuwen F., Feast M., Whitelock P., Yudin B., 1997, MNRAS 287, 955

Whitelock P., Menzies J., Feast M., Marang F., Carter B., Roberts G., Catchpole R., Chapman J., 1994, MNRAS 267, 711

MAXIMUM LIKELIHOOD ESTIMATION OF THE SCALE HEIGHT OF MIRA VARIABLES.

ZSOLT BEREND AND LAJOS G. BALÁZS

Konkoly Observatory, Budapest

Abstract. Based on the DENIS data observed in a strip passing near to the Galactic anticenter (l=182° - 209°, b=$-$2° - 13°), and using photometric selection criteria, we extracted a sample of Mira variables. Assuming a spatial distribution in the form of an exponential disk we estimated the scale height using the Maximum Likelihood technique. The procedure also yielded the confidence interval of the estimated parameter.

1. Introduction

The Mira variables are pulsating red giants with high luminosity, lying on the Asymptotic Giant Branch of the H-R diagram. The DENIS project has important role in their studying because they are brightest in the infrared bands. From DENIS strips a sufficient number of red variables can be extracted for statistical studies. Their spatial distributions can help in understanding the structure of the Galactic Disk.

We have to be acquainted with the scale heights of the populations in order to know their distributions. The aim of this work is to give a Maximum Likelihood estimation of scale height of Mira variables.

2. Maximum Likelihood estimation

The spatial distribution of Mira variables can well be approximated assuming exponential distribution (Wainscoat (5)) in z and R (distance from the Galactic plane and centre, respectively). Consider the following function:

$$L = \tilde{A}(m_1, l_1, b_1; z_0)\,\tilde{A}(m_2, l_2, b_2; z_0) \ldots \tilde{A}(m_n, l_n, b_n; z_0), \qquad (1)$$

135

N. Epchtein (ed.),
The Impact of Near-Infrared Sky Surveys on Galactic and Extragalactic Astronomy, 135-138.
© 1998 *Kluwer Academic Publishers.*

where $\tilde{A}$ is the probability density function of stars

$$\tilde{A}(m, l, b, S) = \frac{A(m, l, b, S)\Psi(m)\Xi(l, b)}{\int_{-\infty}^{\infty} A(m, S)\Psi(m)\Xi(l, b)dmdldb}, \tag{2}$$

A gives the amounts of stars in one square degree depending on distance and direction referring to a given spectral type, m_i, l_i, b_i are the apparent magnitude and galactic coordinates of stars of the sample, $\Psi(m)$, $\Xi(l, b)$ are the truncation functions in m, l, b and z_0 is the exponential scale height of the population, and the unknown parameter of the likelihood function. This function takes its maximum at some $\tilde{z}_0 = \tilde{z}_0(m_i, l_i, b_i)$. This $\tilde{z}_0$ value is considered to be the estimation of the real scale height value. The likelihood function (1) gives the probability density of obtaining the available sample in the course of the sampling of $\tilde{A}(m, l, b; z_0)$. It means, that the z_0 value is estimated by that special $\tilde{z}_0$ value, which if it were the real value of our parameter, then exactly the given sample would the most likely ensue among the possible samples of n elements.

If $\tilde{z}_0$ is the Maximum Likelihood estimation of the z_0 parameter, then assymptotically

$$2(L_{max} - L_0) \approx \chi^2,$$

where L_{max} is the maximum of the likelihood function (1), and L_0 is the value taken at the real parameter, an d χ^2 has the degree of freedom of the parameters estimated.

The first confidential area can be obtained by subtracting 0.5 from the likelihood function at maximum and drawing at that point a horizontal line (see Figure 2). The two extremities of the confidential area will be where this line intersects the curve.

By the process written above we obtain the 1 σ probability of $\tilde{z}_0$. This value can be reduced by increasing the number of elements of the sample.

We have written a programme estimating scale height making use of the outlined process. Firstly we tested the programme with simulated data. We produced the samples with Monte Carlo method, and the estimated scale heights were all consistent to the simulated values within 1 σ.

3. Application to DENIS data

The available sample contains the right ascensions, declinations and I, J, K colours of 25054 stars (Field: l=182° - 209° ; b=−2° - 13°).

In order to extract the Mira variables from this strip we plotted the J-K colour indexes in function of I-K (see Figure 1) using the literature data (Bessel & Brett (1); Feast et al. (2)). 77 Mira type variables have been left after the selection. Miras can be classed into populations according to

Figure 1.   Field: l=182° - 209° ; b=−2° - 13°

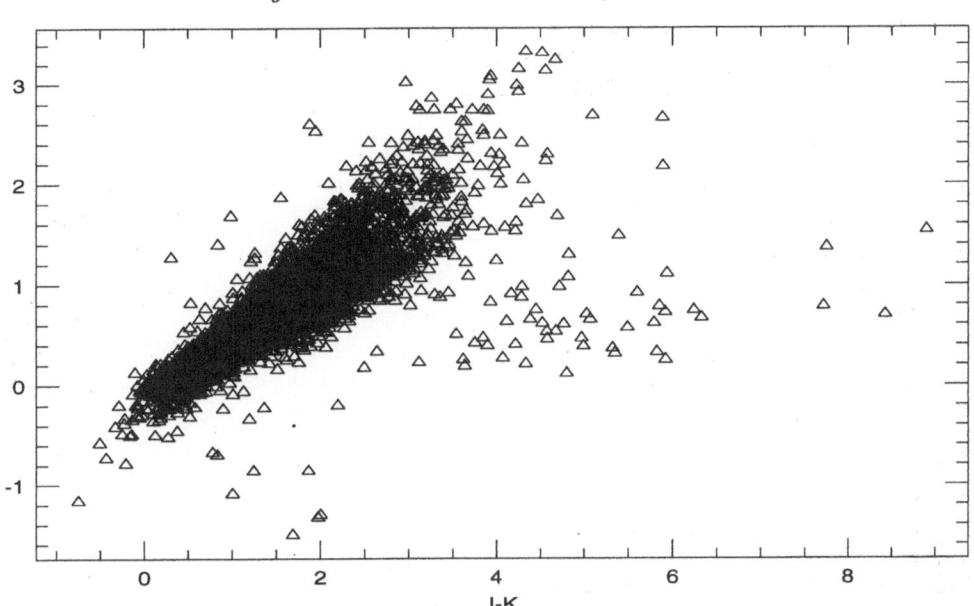

Distribution of the sample in the $J - K$, $I - K$ plane. Mira variables can be found in the range: $3.3 < I - K < 6$ and $1 < J - K < 1.4$.

their periods, and each population represent a different subsystem of the Galactic Disk with different scale height (Jura & Kleinmann (4) ; Hron (3)). However we have not got any information of the periods we only know the apparent magnitudes. Therefore the estimated scale height of this sample is a mean value of the populations. In order to calculate the maximum of the likelihood function we used numerical methods. We increased the z_0 parameter by 10 pc from a chosen initial value, and calculated its likelihood function at every step with numerical integration. The point at which the value of the likelihood function is the highest is considered to be the estimated scale height of the sample.

The likelihood function obtained by this method is in Figure 2. The scale height is <u>510</u> pc and the region surrounded by solid lines represent the 1 σ confidence interval.

Acknowledgements

Special thanks to Stéphanie Ruphy for providing us with the DENIS strip data (private communication 1996).

Figure 2. Likelihood Function

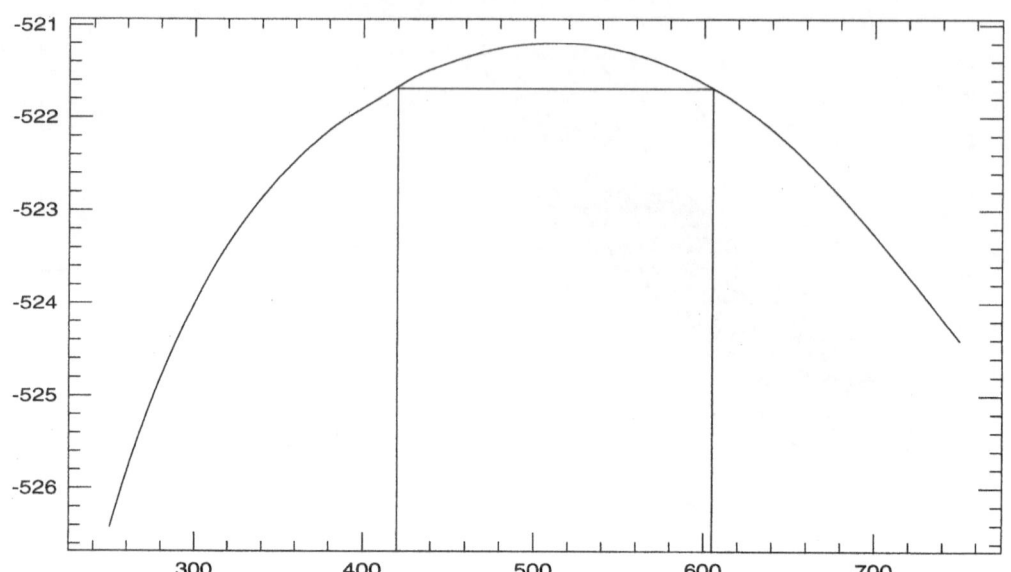

The maximum of the Likelihood Function is at $z_0=510$ pc, the selected region represents the first confidence interval.

References

Bessel, M.S., & Brett, J.M. 1988, PASP, 100, 1134

Feast, M. W., Robertson, B. S. C., Catchpole, R. M., Lloyd Evans, T., Glass, I.S., & Carter, B. S. 1982, MNRAS, 201, 439

Hron, J. 1991, A&A, 252, 583

M. Jura & S. G. Kleinmann : 1992, ApJS 79, 105-121

Wainscoat, R. J., Cohen, M., Volk, K., Walker, H. J, & Schwartz, D.E. 1992, ApJS, 83, 111

IV- Star forming regions and Interstellar medium

2MASS OBSERVATIONS OF MOLECULAR CLOUDS

JOHN M. CARPENTER
California Institute of Technology
Department of Astronomy, MS 105-24, Pasadena, CA 91125

1. Introduction

A major goal of star formation studies is to characterize the conditions and "mode" under which most stars form in our Galaxy (see, e.g. Lada, Strom, & Myers 1993). That is, among the continuum of stellar systems found in molecular clouds, how are most stars created? Are they formed primarily in rare, but densely populated clusters, or among smaller clusters and isolated star forming regions? Do most stars form uniformly in time throughout the lifetime of a molecular cloud, or in sporadic bursts? What causes one portion of a molecular cloud do form rich clusters, but not other regions? How (and if at all) do the properties of the stars (e.g. lifetime of protoplanetary disks, the stellar initial mass function, etc...) vary among the different stellar systems? While answering these questions requires substantial photometric and spectroscopic observations, the first crucial step requires determining the size and spatial distribution of the stellar population within molecular clouds.

Determining whether most stars form in isolation or in clusters has far broader implications than merely establishing the predominant morphological distribution of young stellar objects. First, it establishes the environmental conditions most stars form under, which in turn may influence the resultant stellar properties. For example, the stellar densities observed in some clusters imply that the initial protostars may have undergone substantial gravitational and tidal interactions (Zinnecker, McCaughrean, & Wilking 1993). Such interactions may alter the mass accretion rate onto a star, the lifetime of protoplanetary disks, the final stellar mass, and the formation of binary systems (Ostriker 1994; Kroupa 1995; Price & Podsiadlowski 1995; Jensen, Mathieu, & Fuller 1996; Bonnell et al. 1997). The spatial distribution of stars can also potentially constrain various evolutionary scenarios for molecular clouds. Dense clusters are easily recognized

141

N. Epchtein (ed.),
The Impact of Near-Infrared Sky Surveys on Galactic and Extragalactic Astronomy, 141-153.

as a group of stars forming in a localized region of space. Many clusters are localized in time as well in that they represent "bursts" of star formation where hundreds of stars or more are formed on time scales of less than a few million years (Hillenbrand et al. 1993; Greene & Meyer 1995; Meyer 1996; Hillenbrand 1997). Dynamical considerations also suggest that compact clusters constitute relatively young stellar systems (Lada, Margulis, & Dearborn 1984). Thus a stellar population consisting largely of clusters would suggest that star formation has not taken place over long periods of time, either since the cloud is young, or because star formation has been inhibited over most of the cloud lifetime. On the other hand, a molecular cloud containing primarily a large scale, uniformly distributed population of stars is more difficult to interpret, as such a population may result from numerous isolated star forming events or small, old clusters that have since dispersed over large spatial scales. While these scenarios do not exhaust all possible situations and interpretations, it is clear that the observed spatial distribution of stars provides an important initial step toward establishing how star formation proceeds within molecular clouds.

The primary difficulty in determining the spatial distribution of stars in clouds is identifying young stellar objects actually associated with the cloud amongst the population of unrelated field stars. Classically this has been accomplished by selecting individual stars based on the presence of an emission feature associated with young stellar objects (e.g. $H\alpha$, near-infrared excesses, far-infrared emission, x-rays, etc...). While these methods generally have a high reliability in finding young stellar objects, they have severe selection biases in that these emission features (except x-rays) are generally associated with pre-main-sequence stars surrounded by optically thick accretion disks. Since the evolution of these disks and emission features as a function of time and stellar mass is unknown, it is not possible to infer the extent of the stellar population that has already passed through this phase of stellar evolution and is not detectable by these methods. While x-rays surveys probe chromospheric activity and can detect older stars, these surveys often provide a vastly incomplete census of the stellar population. In the Orion region, for example, the *ROSAT* All Sky Survey detected 850 x-ray sources in a 450 $\deg^2$ area centered on the Orion molecular cloud (Alcalá et al. 1996). The total stellar population in Orion, however, exceeds at least 5000 stars as inferred from near-infrared observations (Lada et al. 1991; Strom, Strom, & Merrill 1993; Ali & DePoy 1995).

As an alternative to definitively identifying each individual star associated with a cloud, the underlying stellar population can be inferred statistically using star counts. This method attempts to establish the stellar population by determining the excess number of stars observed toward the cloud with respect to the field star population. This approach is particularly advantageous for near-infrared wavelength observations. As indicated above, near-infrared observations are often more sensitive to the embedded stellar population than peculiar emission surveys. Further, near-infrared emission generally probes the stellar photosphere (except for heavily accreting objects). Thus the sensitivity of these surveys as a function of stellar mass and age are easier to estimate. While optical observations also possess these advantages, the lower extinction at near-infrared wavelengths have enabled these surveys to probe deep into molecular clouds and discover stars not detectable at optical wavelengths. Thus at the cost of uniquely identifying young stars, near-infrared star count studies can potentially obtain a more complete census of the embedded stellar population.

In the following discussion, I examine the potential that the current generation of large scale near-infrared surveys – and 2MASS in particular – have for extending our understanding of the stellar population in molecular clouds through star count studies. I will begin by reviewing in §2 recent results on characterizing the stellar population within clouds, and indicate the need for surveys such as 2MASS to answer many of the outstanding questions. In §3, I examine the sensitivity of 2MASS to young stars as a function of stellar mass, age, distance, and extinction, and identify regions in the Galaxy that are particularly well suited for star count studies with the 2MASS data base. Details of a star count analysis are considered in §4, and well as some of the numerical methods that hold promise to objectively analyze the data. A summary of my conclusions is provided in §5.

2. Recent Results

Much of our understanding of the distribution of stars within molecular clouds comes from studies of two regions: Taurus and Orion. Taurus has been studied at nearly every available part of the electromagnetic spectrum, and a compilation of these studies shows that the known stellar population in Taurus consists of ~185 stars (Kenyon & Hartmann 1995), at least half of which can be assigned to six small groups of ~15 stars each (Gomez et al. 1993). Interestingly, spectroscopic observations have shown that stars in Taurus generally have ages of $\lesssim 3$ Myr (Kenyon & Hartmann 1995), and that a widespread population of old stars does not appear to be present. Since molecular clouds typically have lifetimes in excess of 10 Myr (Leisawitz, Bash, & Thaddeus 1989; Elmegreen 1991), these results suggest that unless the Taurus region is much younger than the average molecular cloud, star formation has occurred only in its recent history. These conclusions remain controversial, however, as there has been much debate on the completeness of current surveys in sampling the total stellar population in molecular clouds (Feigelson 1996; Briceño et al. 1997).

One complication in generalizing the above results is that most of the molecular mass in the Galaxy is contained in Giant Molecular Clouds (GMCs) that have masses an order of magnitude or more larger than Taurus (Sanders, Scoville, & Solomon 1985). Logically one assumes that most stars form in GMCs as well, although it remains to be demonstrated that this is actually the case. The nearest example of a GMC is the Orion molecular cloud, which like Taurus has been the subject of numerous studies at multiple wavelengths. The most sensitive and systematic of these surveys designed to address the importance of cluster and distributed stellar populations in Orion have been conducted in the near-infrared. These surveys indicate that the stellar population in the northern half of Orion (L1630) is contained almost exclusively in 4 rich, dense clusters (Lada et al. 1991; Li, Evans, & Lada 1997). The southern half of Orion (L1641) contains roughly an equal number of stars, however, they are distributed uniformly throughout the cloud (Strom, Strom, & Merrill 1993) and not in rich dense clusters that characterize the northern half of Orion. The implications of these disparate conclusions concerning the spatial distribution of stars for different parts of the same molecular cloud are unclear. These results might suggest that no dominant mode of star formation exists, and that the evolution of clouds and their star formation content depends strongly on the local conditions. Alternatively, perhaps clusters are the dominant mode of star formation, and the distributed stellar population found in L1641 represents clusters formed long ago but have since dispersed. Note, however, that these two studies have imaged just 4% of the projected surface area of the Orion molecular cloud (Maddalena et al. 1986). Further, neither survey encompasses the Trapezium region, which has a total stellar population that exceeds what has been inferred so far in L1630 and L1641 combined (Ali & DePoy 1995).

Clearly any conclusions concerning the stellar population within molecular clouds remain speculative. Existing surveys often do not cover entire clouds, and the number of clouds that have been observed are so few in number that any generalized statements are not possible. The primary difficulty in this endeavor of course is the inordinate time required to map entire molecular clouds to sensitive levels at near-infrared wavelengths. These limitations, however, will soon be lifted by the 2 Micron All Sky Survey (2MASS), which has begun to scan the entire sky to a 10σ completeness limit of 15.8^m, 15.1^m, and 14.3^m at J, H, and K_s band respectively. The large area coverage, the uniform sensitivity, and the multi-wavelength observations inherent to 2MASS provide powerful tools to infer the underlying stellar population in molecular clouds. In the remainder of this contribution, I investigate the practical aspects involved in conducting star count analyses with the 2MASS data base.

3. Sensitivity to Young Stellar Objects

Before discussing the computational aspects of analyzing star count data, it is important to establish how well 2MASS can probe the underlying stellar population associated with a molecular cloud. The primary factors to consider are *(i)* the sensitivity of 2MASS to young stellar objects as a function of stellar mass, age, distance, and extinction; and *(ii)* the constraints imposed by field stars in inferring the stellar population associated with the cloud. I discuss these issues below and demonstrate that the stellar population can be constrained to interesting limits for clouds in the solar neighborhood. This discussion also serves to underscore subtle aspects that must be considered in the data analysis in order to accurately characterize the stellar population.

3.1. STELLAR AGES, MASSES, AND DISTANCES

Since most stars associated with molecular clouds are in the pre-main-sequence phase of stellar evolution, their brightness will depend sensitively on the stellar age and mass. The range of stellar masses and ages that 2MASS can probe is illustrated in Figure 1 for four well-known star forming regions that span a range of distances from the sun: Taurus (distance = 140 pc), Orion (450 pc), W3 (2200 pc), and W51 (7000 pc). The solid curves in Figure 1 represent K_s band isomagnitudes calculated using the pre-main-sequence evolutionary tracks from D'Antona & Mazzitelli (1994) assuming zero extinction, and the grey shaded region highlights the parameter space that 2MASS can probe at the 10σ K_s band completeness limit (14.3^m). For example, Figure 1 shows that at the distance of Taurus, 2MASS will be able to detect all stars younger than 10^8 years that are now on the cloud surface (i.e. $A_V = 0^m$). Even for visual extinctions as large as 20^m, the 2MASS survey will be able to detect nearly all stars within this age range ($A_K \approx A_V/10$). Given that the age of a typical cloud is thought to be less than 10^8 years (Leisawitz, Bash, & Thaddeus 1989; Elmegreen 1991), 2MASS will be able to detect nearly all stars in Taurus with $A_V < 20^m$ that have formed throughout the cloud lifetime. At the distance of Orion, 2MASS will be able to detect a surface population of stars more massive than 0.4 $M_\odot$ that have formed over the cloud lifetime, and at the distance of W3, a surface population of 1.4 $M_\odot$ stars is readily detectable. Beyond a couple of kiloparsecs, only young, massive stars will be easily detected.

Besides the obvious decrease in sensitivity with increasing distance, Figure 1 also emphasizes that for a given stellar mass, the sensitivity decreases with increasing stellar age. To illustrate the effect time evolution has on the detectability of the underlying stellar population within a cloud, Figure 2 shows the fraction of stars in a model molecular cloud that are brighter

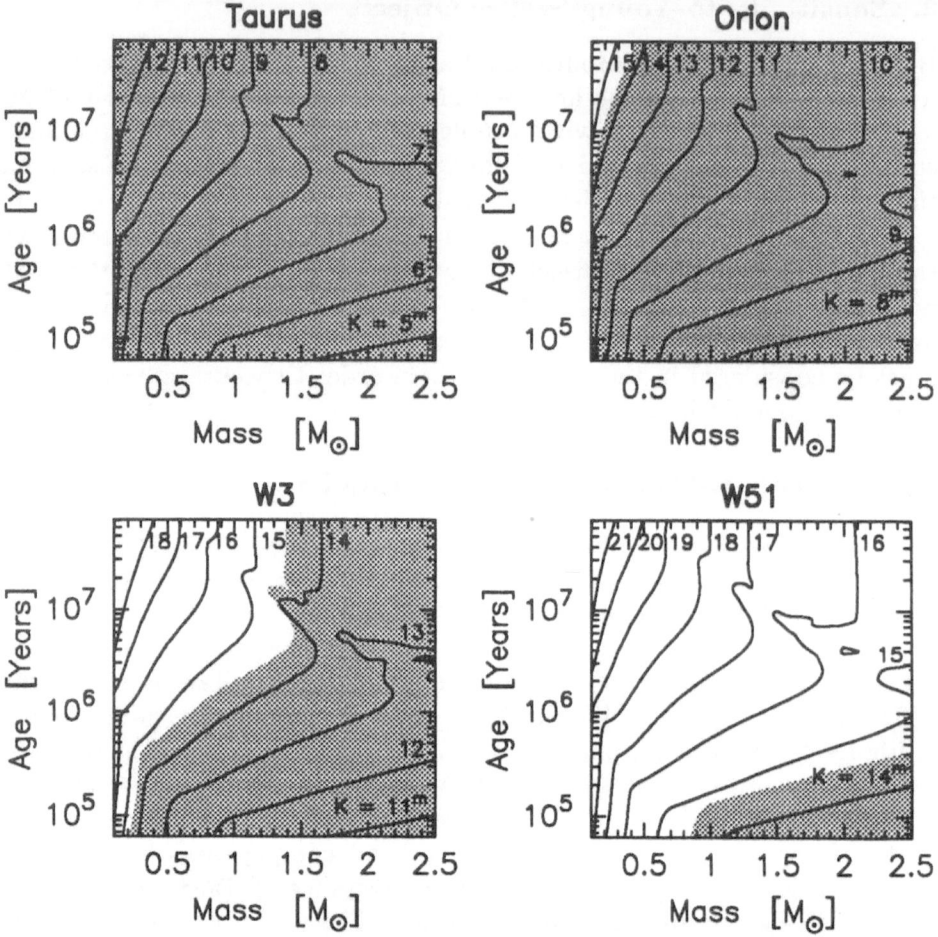

Figure 1. K_s band isomagnitude contours as a function of stellar age and mass for stars at the distance of Taurus (140 pc), Orion (450 pc), W3 (2200 pc), and W51 (7000 pc). The magnitudes were computed using the pre-main-sequence evolution tracks from D'Antona & Mazzitelli (1994; model 1 with CM convection and Alexander opacities) assuming $A_V = 0^m$ and that no excess near-infrared emission is present. The shaded area highlights the parameter space that 2MASS can probe at the 10σ K_s band completeness limit (14.3^m). Since clouds are thought to have ages of less than 10^8 years, 2MASS can detect all stars now on the surface of Taurus that have formed throughout the cloud lifetime. For Orion, a surface population of stars more massive than $0.4\,M_\odot$ that have formed over the cloud lifetime will be detectable. One challenge in inferring the size of the underlying stellar population within a cloud is to take into account the sensitivity of 2MASS to stars of various masses, ages, and extinction.

than the 2MASS K_s band completeness limit assuming that stars have been forming at a constant rate over time with a Miller-Scalo stellar initial mass function (Miller & Scalo 1979). Three different time periods for

the star formation episode are indicated in the figure. At the distance of
Orion, for example, if star formation has occurred only in the past 10 Myr,
then ~90% of the stars will be detectable with 2MASS. (Note, however,
that extinction has been ignored for this illustration.) Only about half of
the stars will be visible though if star formation has taken place over a
time period of 70 Myr. Since a distributed population of stars might be
expected to be older on average than stars found in clusters, the sensitivity functions shown in Figure 1 must be considered when evaluating the
relative significance of cluster and distributed stellar populations.

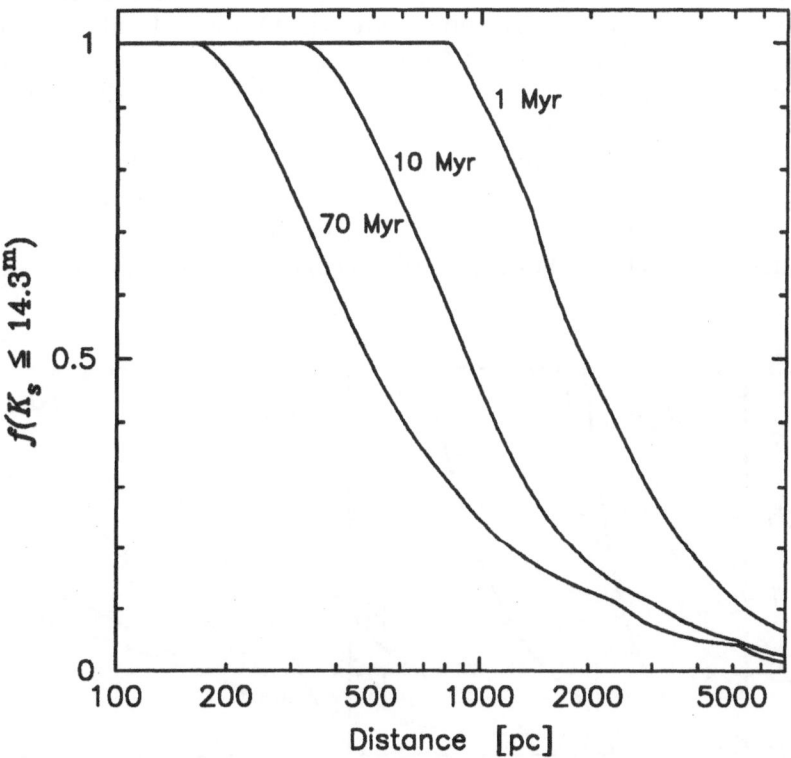

Figure 2.   The solid curves represent the fraction of stars brighter than the 2MASS K_s
band completeness limit (14.3^m) in a model molecular cloud that has been forming stars
at a constant rate over time with a Miller-Scalo initial mass function. The results are
shown for when star formation has occurred constantly over three different time periods:
1 Myr, 10 Myr, and 70 Myr. No extinction has been added to the models, and the stellar
mass range spans from 0.1 $M_\odot$ to 20 $M_\odot$. At a distance of 500 pc, for example, 2MASS
can detect anywhere from 100% of the stars if star formation has persisted for only 1 Myr,
to as low as 48% of the stars if star formation has continued for as long as 70 Myr.

3.2. EXTINCTION

Besides the stellar mass, age, and distance, extinction will have important consequences for determining the visibility of a star. For most regions, extinction will come primarily from dust associated with the parental molecular cloud and not from dust in the general interstellar medium. To illustrate the effect that extinction has on the sensitivity of 2MASS to young stellar objects, Figure 3 shows the lowest mass star that can be detected at J, H, and K_s bands for the 10σ completeness limits as a function of depth (i.e. visual extinction) into a cloud. The stellar magnitudes were computed using the pre-main-sequence evolutionary tracks from D'Antona & Mazzitelli (1994) assuming stellar ages of 1 Myr (left panel) and 20 Myr (right panel) and a distance of 450 pc (corresponding to the Orion molecular cloud). Figure 3 shows that for an age of 1 Myr, the J, H, and K_s band surveys are sensitive to 0.1 $M_\odot$ stars (and the more massive brown dwarfs) on the cloud surface. The J band surveys quickly lose sensitivity to the low mass stars as the extinction increases, and for $A_V > 10^m$, K_s band is the most sensitive probe of embedded stars regardless of the age. Even for $A_V = 30^m$, the 2MASS K_s band survey will detect 1 Myr stars more massive than 0.4 $M_\odot$ at the distance of Orion. For $A_V < 10^m$, H band is generally more sensitive than K_s band.

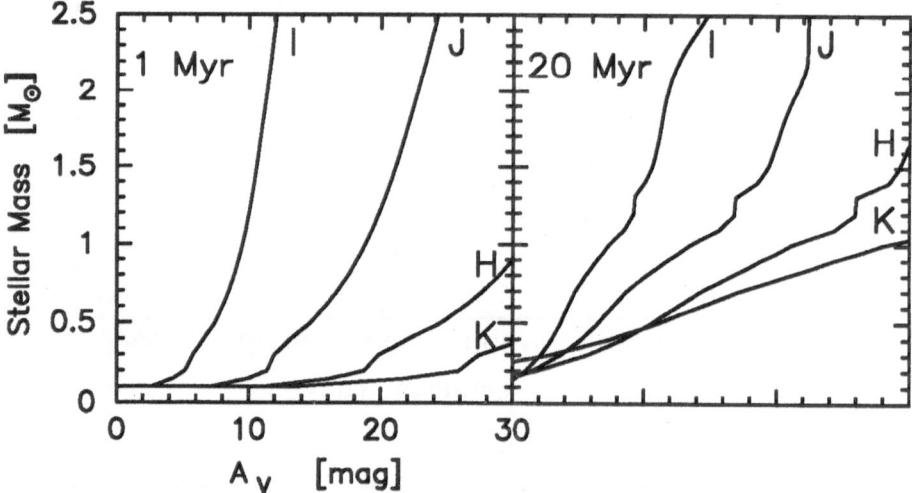

Figure 3. The lowest mass pre-main-sequence star that can be detected at various depths into the cloud (as measured by visual extinction) at the distance of Orion (450 pc) for the 10σ completeness limits of 2MASS at J (15.8^m), H (15.1^m), and K_s (14.3^m) bands. The magnitudes were computed using the pre-main-sequence evolutionary tracks from D'Antona & Mazzitelli (1994). The left panel shows the results for stars with an age of 1 Myr and the right panel for an age of 20 Myr. These curves show that the 2MASS H band survey is generally the most sensitive probe of young stellar objects for $A_V < 10^m$, and the 2MASS K_s band survey is more sensitive for $A_V > 10^m$.

For actual molecular clouds, the visual extinction for a random line of sight is typically $A_V \lesssim 5^m$ and for a star forming dense core, $A_V \gtrsim 10^m$ (Carpenter, Snell, & Schloerb 1995; Heyer, Carpenter, & Ladd 1996). Thus the most sensitive probe of stars randomly situated throughout a molecular cloud are the 2MASS H band observations. These data are particularly important for tracing a widespread distributed stellar population since many of these stars are expected to be old and have low visual extinctions. In dense cores, however, the 2MASS K_s band data are best used to detect embedded objects. Thus while I emphasize analysis of the K_s band data in this discussion, the most complete analysis of the stellar population will actually be obtained from an aggregate star count map constructed from the J, H, and K_s band 2MASS data.

3.3. FIELD STARS

The calculations presented above demonstrate that 2MASS is quite sensitive to young stellar objects in nearby molecular clouds within ~ 1 kpc of the sun (see Fig. 1). However, since the stellar population associated with a cloud will be inferred from star counts (see §4 for details), an additional criteria is that the number of stars associated with the cloud significantly exceeds the random fluctuations among the field star population. These constraints are particularly important for any uniformly distributed stellar population since these stars may be significant in number but low in surface density. The number of field stars (N_f) projected against a circular cloud with angular radius θ_c immersed in a field star backdrop with surface density D_f is $N_f = D_f \pi \theta_c^2$. For Poisson statistics, the minimum *detectable* distributed population associated with the cloud (N_{min}) is

$$N_{min} \propto \sqrt{N_f} \propto \sqrt{D_f}\, \frac{R_c}{d_c},$$

where R_c is the cloud radius and d_c the cloud distance. As an example, the Orion molecular cloud has a projected surface area of ~ 48 deg^2. The estimated field star surface density is ~ 0.35 stars arcmin^{-2} for $K_s \leq 14.3^m$ (Jarrett 1992), implying that the 3σ minimum detectable stellar population is ~ 750 stars for an assumed distance of 450 pc. Since this limit is less than the number of stars found in the various clusters in Orion (Lada et al. 1991; Ali & DePoy 1995), interesting limits on the relative importance of the distributed and clustered populations can indeed be obtained for Orion and nearby clouds. Note that this analysis only considers stars brighter than the photometric completeness limit. As emphasized in Figure 1, old, low mass stars may be present at fainter magnitudes. The number of such stars will depend on the cloud distance and frequency distribution of stellar ages and masses, and is necessarily model dependent.

4. Data Analysis

It was shown in §3 that 2MASS will be able to detect a substantial fraction of the stellar population in nearby molecular clouds and establish the relative contributions of clustered and distributed stellar populations to interesting limits. Since the methods to actually extract this information from the star count data are diverse, I sketch only a general outline here that attempts to analyze the data in a self-consistent manner. In addition, I briefly discuss some of the numerical methods that can be used to objectively identify stellar clusters.

4.1. STAR COUNTS

The observed star counts toward a molecular cloud consists of three components: *(i)* field stars foreground to the cloud; *(ii)* background field stars reddened primarily by dust associated with the cloud; and *(iii)* stars embedded in the cloud. The cloud stellar population can therefore be obtained by subtracting the foreground stars and reddened background objects from the observed star counts. The practical difficulty in this procedure is that near-infrared colors and magnitudes cannot completely distinguish field stars from the stellar population associated with the cloud. Stars with red near-infrared colors can either by reddened background objects or stars embedded within the cloud, and unreddened stars may either be foreground objects, young stars on the cloud surface, or background stars appearing through holes in the cloud. While stars with a near-infrared excess can be unambiguously assigned to the cloud, such a selection criteria only identifies objects surrounded by optically thick accretion disks, which may represent a small fraction of the total stellar population. Also, many stars will be detected at only one or two wavelengths, further limiting the information that can be inferred from the colors and magnitudes.

Rather than attempt to definitely identify each young stellar object, the stellar population associated with a cloud can be statistically determined by subtracting the expected distribution of field stars from the observed star counts. The systematic and random angular variations in the field star surface density can be established quite accurately from the 2MASS observations of the regions surrounding the molecular cloud. A semi-empirical model of the background field star population can then be constructed by using Galactic star count models to estimate the fraction of the observed field stars that are likely background objects. Star count models already provide an accurate description of the near-infrared sky (e.g. Bahcall & Soneira 1984; Garwood & Jones 1987; Jarrett 1992), and they will be even further refined with the extensive 2MASS data base. Next, the semi-empirical background field stellar population needs to be reddened using an extinction map of the cloud. Such maps are now readily obtained from molecular line observations, most notably CO and its isotopes.[1] While it is well established that molecular clouds contain substructure down to the smallest observable spatial scales, the high resolution, fully sampled maps now available resolve the large scale filaments, shells, and sheets of gas within molecular clouds (e.g. Bally et al. 1987; Carpenter, Snell, & Schloerb 1995) and accurately trace large scale extinction variations. Thus after subtracting off the model distribution of unreddened foreground field stars and reddened background field stars, a two dimensional map of the stellar population associated with the cloud is obtained.

4.2. CLUSTER IDENTIFICATION

In many respects, identifying clusters is one of the more difficult parts of the analysis since a cluster is not well defined mathematically. In addition, a cluster in one cloud may not be classified as such in another cloud if, for example, the field star surface density is larger. By most definitions, clusters represent a "significant" enhancement in the stellar surface density with respect to the fluctuations in the field star surface density, and the distributed population consists of stars not found in clusters. Most star formation studies have identified clusters by choosing a stellar surface density threshold in a binned star count map. The difficulty with this analysis is that the surface density threshold and the binning cell size are arbitrary parameters biased against finding compact clusters smaller than the cell size, extended low density regions, and clusters not conforming to the adopted axes for the binned star count map.

Recently, non-parametric algorithms (e.g. wavelets, maximum penalized likelihood estimators, kernel density estimation) have been developed and applied to astronomical problems that avoid arbitrary binning of data and provide a more objective measure of the underlying density distribution (Silverman 1986; Scott 1992; Pisani 1996; Fadda, Slezak, & Bijaoui 1997). An illustration of adaptive kernel density estimation is shown in Figure 4 for a simulated stellar population consisting of 5 clusters superimposed on a random distribution of field stars. In this technique, each star is represented by a kernel function (e.g. a gaussian – the results do not depend sensitively on the actual kernel used as long as it has rapidly falling tails). The kernel size for any individual star is computed based on the local stellar density, and the density estimate for the total stellar population is given by the sum of the individual kernels. This procedure contains a number

---

[1]The star counts cannot be used to determine the extinction (e.g. Lada et al. 1994) since this method assumes that none of the stars are associated with the cloud, which is exactly the stellar population that one is attempting to measure.

of highly desirable features for identifying clusters: *(i)* the method is non-parametric and requires no binning of the data; *(ii)* the individual kernel sizes are computed objectively by minimizing the mean integrated square error of the difference between the kernel density estimate and the observed data (Silverman 1986); *(iii)* the probability that each cluster could result from a random distribution of objects can be readily computed (Materne 1979; Pisani 1996); *(iv)* extension of this technique to n-dimensions (e.g. right ascension, declination, magnitude, etc..) is straight forward. The best manner in which to employ these techniques still needs to be explored, but preliminary results are encouraging.

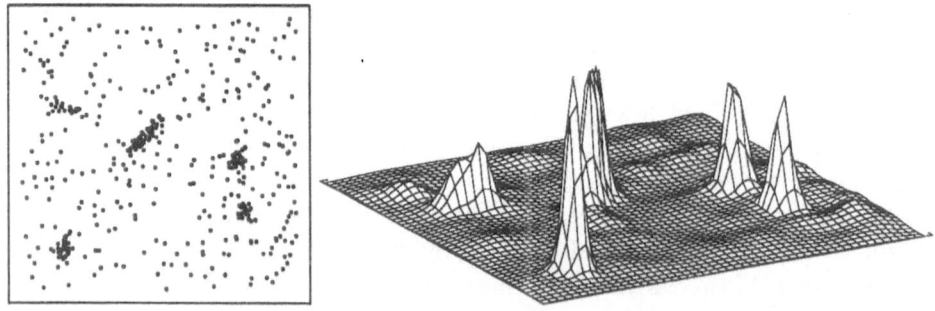

Figure 4. Left panel: The spatial distribution of a simulated stellar population containing 5 clusters superimposed on a random distribution of field stars. *Right panel:* Adaptive kernel density estimate of the model stellar population shown in the left panel (viewed at an angle for clarity). The five clusters are evident as peaks in the density estimate. The number of stars inferred for these clusters agree rather well with the input parameters for the model simulation.

5. Summary

Throughout this discussion I hoped that I have conveyed the excitement and promise that 2MASS provides for investigations into the stellar population of molecular clouds. For nearby objects in particular, 2MASS will be able to determine the distribution of stars that have formed over nearly the entire expected lifetime of molecular clouds. The biggest improvements over existing observations is that for the first time, our understanding of the stellar population in star forming regions will be based on studies of entire ensembles of entire molecular clouds encompassing of range of environments and masses. I look forward to presenting the results of this research at the next Euro conference!

References

Alcalá, J. M. et al. 1996, A&AS, 119, 7

Ali, B. A., & DePoy, D. L. 1995, AJ, 109, 709

Bahcall, J. N., & Soneira, R. M. 1984, ApJS, 55, 67

Bally, J., Langer, W. D., Stark, A. A., & Wilson, R. W. 1987, ApJ, 312, L45

Bonnell, I. A., Bate, M. R., Clarke, C. J., & Pringle, J. E. 1997, MNRAS, 285, 201

Briceño, C., Hartmann, L. W., Stauffer, J. R., Gagné, M., Stern, R. A., & Caillault, J. P. 1997, AJ, 113, 740

Carpenter, J. M., Snell, R. L., & Schloerb, F. P. 1995, ApJ, 445, 246

D'Antona, F., & Mazzitelli, I. 1994, ApJS, 90, 467

Elmegreen, B. G. 1991, in *The Physics of Star Formation*, eds. C. J. Lada & N. Kylafis, (Dordrecht:Kluwer), 35

Fadda, D., Slezak, E., & Bijaoui, A. 1997, A&A, in press

Feigelson, E. D. 1996, ApJ, 468, 306

Garwood, R., & Jones, T. J. 1987, PASP, 99, 453

Gomez, M., Hartmann, L., Kenyon, S. J., & Hewett, R. 1993, AJ, 105, 1927

Greene, T. P., & Meyer, M. R. ApJ, 450, 233

Heyer, M. H., Carpenter, J. M., & Ladd, E. F. 1996, ApJ, 463, 630

Hillenbrand, L. A. 1997, AJ, 113, 1733

Hillenbrand, L. A., Massey, P., Strom, S. E., & Merrill, K. M. 1993, AJ, 106, 1906

Jarrett, T. H. 1992, Ph.D. thesis, University of Massachusetts–Amherst

Jensen, E. L. N., Mathieu, R. D., & Fuller, G. A. 1996, ApJ, 458, 312

Kenyon, S. J., & Hartmann, L. 1995, ApJS, 101, 117

Kroupa, P. 1995, MNRAS, 277, 1522

Lada, C. J., Margulis, M., & Dearborn, D. 1984, ApJ, 285, 141

Lada, C. J, Lada, E. A., Clemens, D. P., & Bally, J. 1994, ApJ, 429, 694

Lada, E. A., DePoy, D. L., Evans, N. J. II, & Gatley, I. 1991, ApJ, 371, 171

Lada, E. A., Strom, K. M., & & Myers, P. C. 1993, in Protostars and Planets III, ed. E. H. Levy & J. Lunine (Tucson: University of Arizona Press), 245

Leisawitz, D., Bash, F. N., & Thaddeus, P. 1989, ApJS, 70, 731

Li, W., Evans, N. J. II, & Lada, E. A. 1997, ApJ, in press

Maddalena, R. J., Morris, M., Moscowitz, J., & Thaddeus, P. 1986, ApJ, 303, 375

Materne, J. 1979, A&A, 74, 235

Meyer, M. R. 1996, Ph.D. thesis, University of Massachusetts

Miller, G. E., & Scalo, J. M. 1979, ApJS, 41, 513

Ostriker, E. C. 1994, ApJ, 424, 292

Pisani, A. 1996, MNRAS, 278, 697

Price, N. M., & Podsiadlowski, PH. 1995, MNRAS, 273, 1041

Sanders, D. B., Scoville, N. Z., & Solomon, P. M. 1985, ApJ, 289, 373

Scott, D. W. 1992, Multivariate Density Estimation, (New York: Wiley)

Silverman, B. W. 1996, Density Estimation for Statistics and Data Analysis, (London: Chapman & Hall)

Strom, K. M., Strom, S. E., & Merrill, K. M. 1993, ApJ, 412, 233

Zinnecker, H., McCaughrean, M. J., & Wilking, B. A. 1993, in Protostars and Planets III, eds. E. H. Levy & J. I. Lunine, (Tucson: University of Arizona Press), 429

STAR FORMING REGIONS AND NEAR-IR SURVEYS

T. MONTMERLE
Service d'Astrophysique, CEA/DAPNIA/SAp
Centre d'Etudes de Saclay
91191 Gif-sur-Yvette Cedex
France
montmerle@cea.fr

Abstract.
The traditional way to find young, low-mass stars has been to do deeper and deeper near-IR surveys of molecular clouds, and to look for objects with excess IR flux over a blackbody photosphere. We thus expect to find many low-luminosity young objects with surveys such as DENIS and 2MASS. However, taken in isolation, such surveys cannot be conclusive in two cases: (i) young objects without near-IR excess (such as the "weak-line T Tauri stars": WTTS); (ii) background cool main-sequence stars or red giants, which dominate the faint population of IR sources. One of the best ways to select young stars without IR excess is to use their X-ray emission properties. As X-ray observations done with ROSAT show, many WTTS are present at a distance from star-forming regions. The spectroscopic determination of their age then allows in particular to trace the past history of star formation of these regions. In this context, I will also discuss the new possibilities offered by the next generation of X-ray satellites, AXAF and XMM

N. Epchtein (ed.),
The Impact of Near-Infrared Sky Surveys on Galactic and Extragalactic Astronomy, 155.
© 1998 *Kluwer Academic Publishers.*

A STUDY OF EXTINCTION AND STAR FORMATION IN THE CHAMAELEON I CLOUD WITH DENIS

L. CAMBRESY

Observatoire de Paris, DESPA, F-92195 Meudon Cedex, France

Abstract.

I present the first massive star count in the J band (1.25 μm) provided by DENIS in the Chamaeleon I dark cloud. These data are used to derive a high resolution map (2′) of an area of 1.5° × 3° around the centre of the cloud using an original processing which involves an adaptive grid for counting and a wavelet decomposition for noise filtering of the extinction map. The maximum extinction in this cloud is found to be 10 Av, using a normal law. Preliminary results on the study of the star formation improve our knowledge about the stellar population of the cloud. In particular, the K_s luminosity function allows the estimation of the age of the stellar population of the cloud.

1. Introduction

The Chamaeleon I cloud is the most obscured region of the Chamaeleon dust-molecular complex. Near infrared (1-2 μm) star counts are more appropriate to probe regions where $A_V \gtrsim 4$ since an extinction of 10 visual magnitudes drops to only ≈ 3 magnitudes in the J band at 1.25 μm. The aim of this paper is to investigate in detail the extinction toward the Cha I cloud using this new wealth of data.

The Cha I dark cloud is located at $b = -16°$ and its current distance estimate is 140 pc (Whittet et al., 1987). Its high galactic latitude implies a small number density of background stars which limits the spatial resolution of the extinction estimation, but, on the other hand, the probability of crossing several clouds on the line of sight is low.

N. Epchtein (ed.),
The Impact of Near-Infrared Sky Surveys on Galactic and Extragalactic Astronomy, 157-163.

2. Star count method

Usually, the extinction is evaluated by comparison of star counts in the absorbed region and a nearby area assumed to be free of obscuration (Wolf diagram method). Star counts are performed by adding up the stars up to a given magnitude (or in a given magnitude range, e.g., $\pm\frac{1}{2}$) within a grid of fixed squares. The step of the grid is a compromise between the stellar density and the spatial resolution. In other words, the spatial resolution is underestimated wherever the extinction is low, while in highly obscured areas, the content of several cells must be merged, in order to pick up enough stars. Moreover the poissonnian error resulting from star counts depends on the extinction.

I have developed a new method to investigate the extinction across a cloud which consists in replacing usual star counts by an estimation of the local projected star density obtained by measuring the mean distance of the x nearest stars. The most important advantage of this method is to match the local extinction: it corresponds to a star count with adaptable square size (Fig. 1).

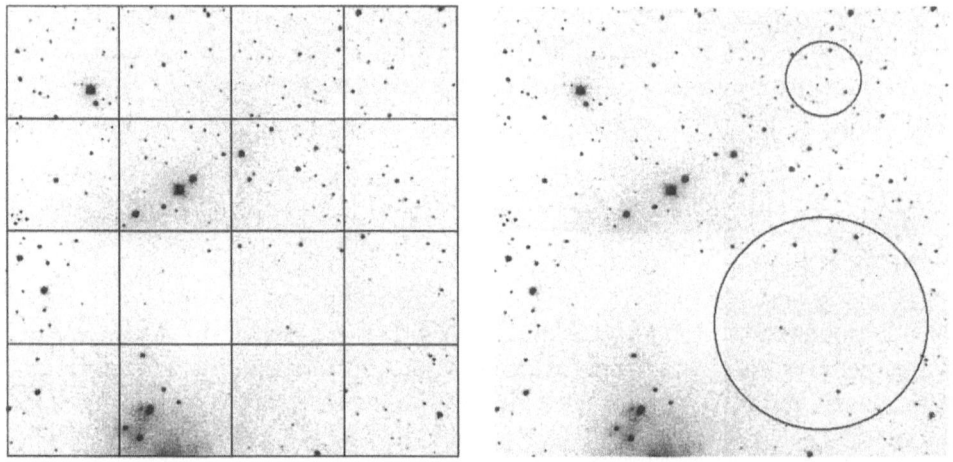

Figure 1.   Classical (left) and adaptive (right) star counts method

Another very interesting advantage of the method is to provide a map with white noise. Therefore, we can simply estimate the noise by computing the standard deviation σ of the mean distance on a part of the map with no signal.

I obtain a map where each point represents the square root of the local density. The extinction is then easily derived by the relation :

$$A_\lambda = \frac{1}{a} \log \left(\frac{\overline{d_{\mathrm{cl}}}}{\overline{d_{\mathrm{cp}}}} \right)^2 \qquad (1)$$

where a is defined by :

$$\log \left(\overline{d_{\mathrm{cp}}} \right)^{-2} = a \times m_\lambda + b \qquad (2)$$

where m_λ is the magnitude, $\overline{d_{\mathrm{cl}}}$ the mean distance of the x nearest stars in the cloud and $\overline{d_{\mathrm{cp}}}$ the mean distance of the x nearest stars in a comparison field supposed unobscured. I verified that the relation (2) is correct up to $J = 16$, i.e. our limit of completeness. Then we convert A_J into visual extinction using the extinction law of Cardelli et al. (1989) for $R_V = \frac{A_V}{E_{B-V}} = 3.1$. So $\frac{A_J}{A_V} = 0.282$.

I can consider the map as a digitized *image* which allows to use current technics of image processing such as the wavelet transform to restore the image and to filter the noise (Starck & Murtagh, 1994). I apply the *à trous* wavelet transform algorithm to split-off the image into 4 wavelet planes (Fig. 2). The decomposition is made by convolving the image by a low–pass filtering matrix. The difference between the original image and the result of the first convolution gives the first plane of the wavelet transform which corresponds to the high frequency plane. Further iterations of this process provide the 4 wavelet planes and the final smooth plane.

Thus, we can use the high frequency plane to identify aberrant points and remove them in the final image in order to eliminate their contribution in all the planes, by replacing the bad pixels by the average of the surrounding 8 pixels. The fact that this plane contains only noise indicates that the extinction measurement are over sampled.

Lastly, we filter each wavelet plane at $3\sigma_i$, where σ_i corresponds to the standard deviation of the plane i. It could be estimated in a part of the image without signal, or it could be simulated. The simulation is very easy since the number of stars in each cells is fixed to x. So, a Poisson noise is generated and we replace in the equation (1) the expression $\left(\frac{\overline{d_{\mathrm{cl}}}}{\overline{d_{\mathrm{cp}}}} \right)^2$ by $\left(\frac{x}{P(x)} \right)$ where $P(x)$ is a Poisson distribution of mean x. I apply the wavelet transform to this simulated noise and we obtain the σ_i for each plane.

3. Results

The final result is the extinction map presented in Fig. 3. This map results from the recombination of the 4 wavelet planes.

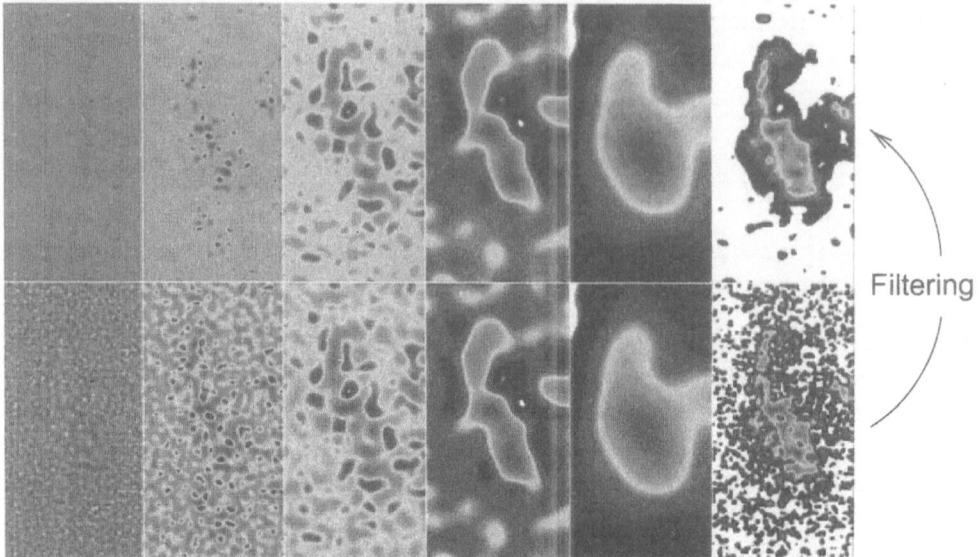

Filtering

Figure 2. Wavelet decomposition of the extinction map. The five first images represent the wavelet planes and the last smooth plan. The lower right image is the raw map, the upper right image is the filtered extinction map

Comparison between near–infrared extinction and IRAS 100 μm emission has been discussed in Cambrésy et al. (1997). In particular, the excellent correlation suggests that the J extinction and the *cold* 100 μm emission have the same origin, a result in agreement with the Désert at al. (1990) dust model which shows that the 100 μm emission and the near infrared extinction are both caused by big grains. The relation between IRAS emission near–infrared extinction is constrained by two parameters : the evolution of the temperature inside the cloud and the extinction efficiency ratio $\frac{Q_e(100\mu m)}{Q_e(J)}$.

This accurate mapping of the extinction is very useful to study the star formation in the cloud. Assuming that the majority of the stars are behind the cloud, we can use the extinction map to deredden them. According to the *Besançon* model (Robin and Crézé, 1986), 98% of the sample consists of background stars. The dereddening applied to the associated members of the cloud is an upper limit. Then we represent each star in a colour–magnitude diagram (K_s vs $J - K_s$) and we select stars which are separated from the main sequence by a distance corresponding to 8 magnitudes of visual extinction, at least. After removing the brightest sources which probably correspond to background giant stars and faintest sources for which the photometric errors are too large, we obtain a sample of 58 good candidates of young stars (Cambrésy et al., in preparation). The number of known

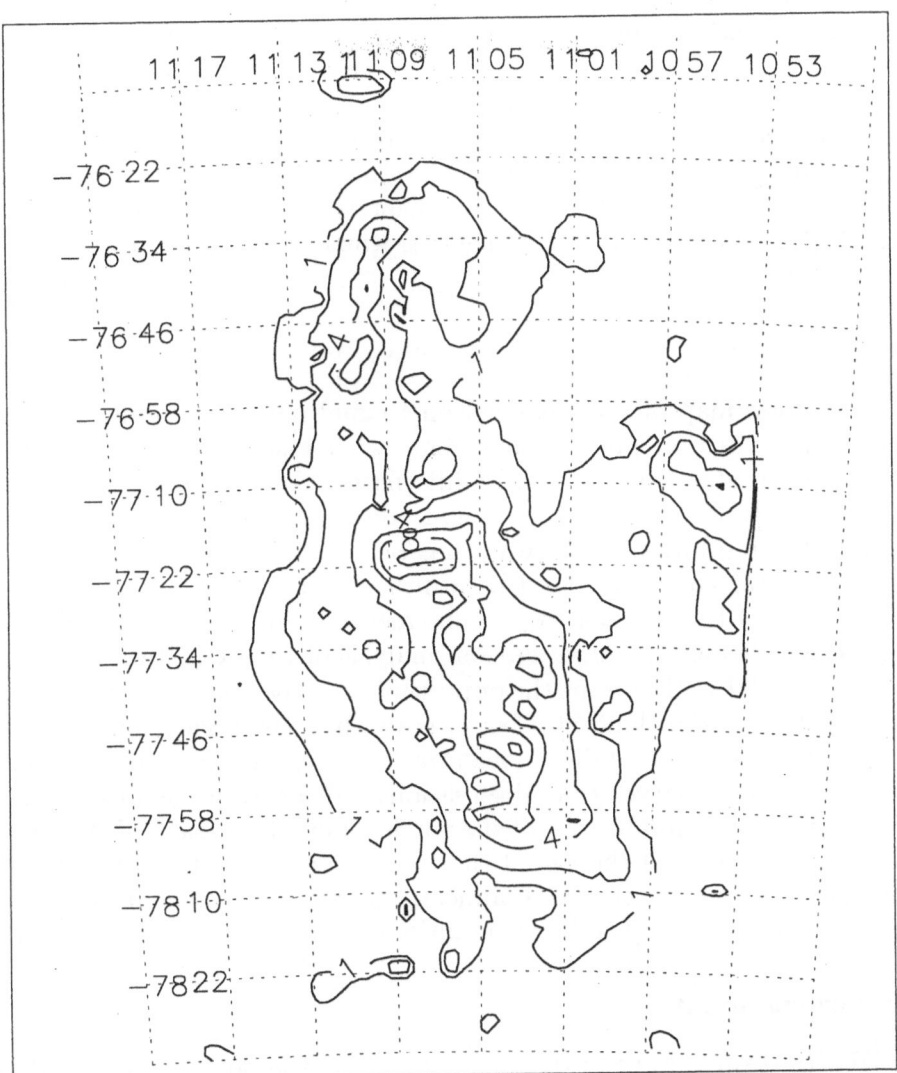

Figure 3. Extinction map derived from *J* band stellar counts. Isocontours correspond to 1, 2, 4, 6, 8 and 10 magnitudes of visual extinction

T–Tauri stars in the Chamaeleon I cloud is 125 (Feigelson and Kriss 1989; Schwartz 1991; Gauvin and Strom 1992; Hartigan 1993; Lawson et al. 1996), so after confirmation we would increase significantly the size of this sample. A spectroscopic diagnosis is now required to confirm the nature of these objects, but their concentration near the cores of the cloud is in favour of their association with the cloud. Another explanation could be the presence of small highly extinguished globules in the cloud which would strongly affect the colour of background stars. These globubes should have a size smaller than 1′ to escape detection in the star count process and thus, extinction should be greater than 15 visual magnitudes. Finally, the selection board only on a colour criterion can confuse unresolved binaries. A forthcoming work will present a thorough study of the star formation in the Chamaeleon I cloud. Preliminary results based on theoretical model by D'Antona and Mazzitelli (1994) show that most sources would have an age ranging from $4\,10^5$ to $3\,10^6$ years.

4. Conclusion

The extinction map of the Chamaeleon I cloud has been significantly improved for extinction greater than $4A_V$ with respect to previous maps obtained from star counts on Schmidt plates. Four distinct maxima are detected and we reach 10 visual magnitudes of extinction without degradation of the resolution. This result has been obtained both by exploiting the massive star counts in the J band provided by DENIS, and by applying a variant of the classical star count method which is adapted to large variations of extinction and a wavelet analysis of the extinction map.

Moreover, DENIS gives us the opportunity to investigate the young stellar population of the cloud at a larger scale than the earlier investigations which were limited to small regions around the visible reflection nebulae. This homogeneous large scale observations will be very useful to investigate the star formation history of the whole cloud. The depth of the survey allows an estimation of the age of the stellar population thanks to the delimitation of a peak in the K_s luminosity function of the cloud members. The age estimate for the stellar population of the cloud is about 10^6 years.

Acknowledgements

The DENIS team is warmly thanked for making the observations available for scientific analysis.

References

Boulanger F., Bronfman L., Dame T.M., and Thaddeus P. (1997), *submitted*

Cambrésy L., Epchtein N., Copet E., de Batz B., Kimeswenger S., Le Bertre T., Rouan D. and Tiphène D. (1997), *A&A*, **324**, pp.L5-L8

Cardelli J.A., Clayton C. and Mathis J.S. (1989), *ApJ*, **345**, pp.245

D'Antona, F. and Mazzitelli, I. (1994), *ApJS* **90**, 467

Désert F.-X., Boulanger F. and Puget J.L. (1990), *A&A*, **237**, pp.215

Feigelson, E. and Kriss, G. (1989), *ApJ*, **338**, pp. 262

Gauvin, L. and Strom, K. (1992), *ApJ*, **385**, 217

Gregorio Hetem J.C, Sanzovo G.C. and Lépine J.R. (1988), *A&AS*, **76**, pp.347

Hartigan, P. (1993), *AJ*, **105**, 1511

Jones T.J., Hyland A.R., Harvey P.M., Wilking B.A. and Joy M. (1985), *AJ*, **90**, pp.1191

Laureijs R.J., Clarck F.O. and Prusti T. (1991), *ApJ*, **185**, pp.372

Lawson, W. A., Feigelson, E. D., and Huenemoerder, D. P. (1996), *MNRAS*, **280**, 1071

Robin, A. and Crézé, M. (1986), *A&A* **157**, 71

Schwartz R.D. and Henize K.G. (1983), *AJ*, **88**, pp.1665

Schwartz, R. (1991), in B. Reipurth (ed.), *Scientific Report*, No. **11**, p. 93, ESO

Starck J.L. and Murtagh F. (1994), *A&A*, **288**, pp.342

Whittet D.C.B., Kirrane T.M., Kilkenny D., Oates A.P. and Watsonf G. (1987), *MNRAS*, **224**, pp.497

DENIS AND ISOCAM OBSERVATIONS OF CHAMAELEON I CLOUD

P. PERSI
Istituto Astrofisica Spaziale, CNR, CP.67,00044 Frascati, Italy

E. COPET
Observatoire de Paris, DESPA, F-92195 Meudon Cedex, France

AND

A.A. KAAS
Stockolm Observatory, S-133 36 Saltsjöbaden, Swenden

Abstract. We present the results of the observations of the Chamaeleon I molecular cloud, in DENIS near-infrared (I, J and K_s) and ISOCAM mid-infrared (LW2, LW3).

1. Observations

1.1. ISOCAM DATA

The Chamaelon cloud has been surveyed with the ISOCAM instrument (Césarsky et al, 1996) in the two broad band filters LW2 (6.75 μm) and LW3 (15 μm). The PFOV was 6 arsec and $T_{int} = 2.1$ seconds, but for some sub-regions wich contain bright sources, $T_{int} = 0.28$ seconds. Then, the dataconsist of 5 rasters map but not uniformely limited in sensitivity.

303 sources have been detected in a region of approximately 1964 sq. arcmin in Cham I including the two B9 stars HD 97300 (North) and HD 97048 (South) which is associated with the bright reflection nebulae Ced 112 and Ced 111. 64% of the sources have been detected only in LW2, 33% show emission in both LW2 and LW3, while approximately 3% have been found only in LW3. The spatial distribution of the sources is shown in Fig. 1. We could detect two clusters, a cluster of very red ISOCAM sources (m(6.75)-m(15) > 1.2) is lovated in the northern part of the cloud (Cham I North), and the other around the star HD 97048. The superimposition on the extinction map of the cloud (Cambrésy et al, 1997) shows that both

165

N. Epchtein (ed.),
The Impact of Near-Infrared Sky Surveys on Galactic and Extragalactic Astronomy, 165-170.
© *1998 Kluwer Academic Publishers.*

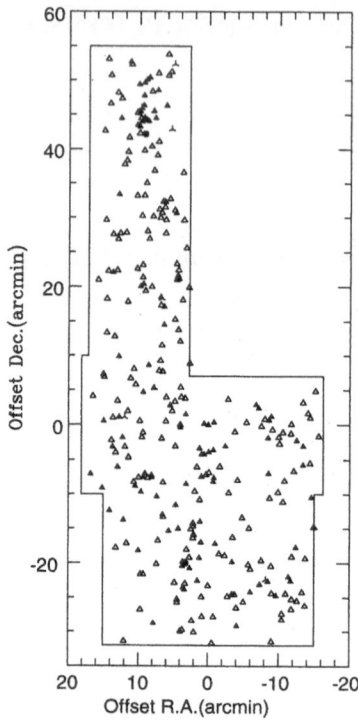

Figure 1.   Spatial distribution of the 303 sources found with ISOCAM. The symbols
are: open triangle, source detected only in LW2; filled circle, detected only in LW3; filled
triangle sources detected in both bands. The (0,0) correspond to R.A.(1950)= 11h05m34s
Dec(1950)=-77d02'

YSO clusters are located in the densest part of the cloud.

The comparaison of the LW2 and LW3 luminosity functions with the
"SKY Model" (Cohen, private communication) shows a cutoff at about
2 mJy for LW2 and 5 mJy for LW3, corresponding roughly to the limits
of sensitivity of the ISOCAM survey. In addition, the observed LW3 star
count shows an excess and a peak at about 77 mJy with respect to the SKY
model, indicating that in Cham I YSO's population is brighter probably
than 8th magnitudes (12 mJy) at 15 μm.

1.2. DENIS DATA

The DENIS (Epchtein et al, this issue) images were acquired in survey
mode, with T_{int} = 9 seconds in I (0.8 μm), T_{int} = 9×1 seconds in both

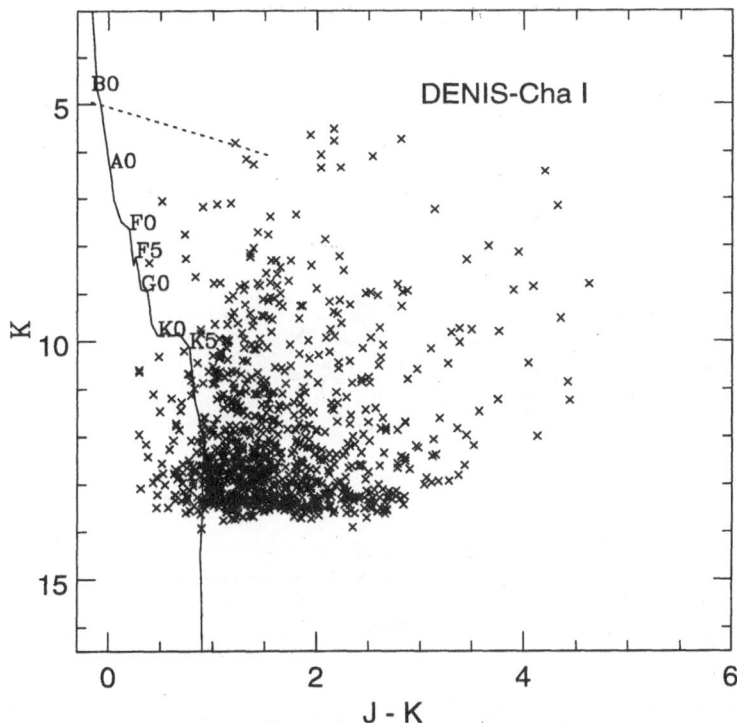

Figure 2. K_s, (J-K_s) diagram the 1172 sources detected by DENIS. the main-sequence is draw in solid line.The dashed line indicate the reddening vector for A_V=10

J (1.25 μm) and K_s(2.16 μm). 13 DENIS "strips" were used to cover the full ChamI cloud. Limiting magnitudes are 18, 16 and 14 at 3 σ level in I, J and $K_s$bands, respectively. 1172 sources have been detected by DENIS in the $K_s$band at a limit of 14 mag in the same ISOCAM surveyed region, 975 have been found also in J.

The color-magnitude K_s, (J-K_s) of the DENIS sources is shown in Fig. 2. Most of the sources common to DENIS and ISOCAM are located in the region (J-K_s)> 1.5 and K_s< 13 of the diagram. The color magnitude diagram in a reference area (not shown) present a concentration in the K=13.5-14, J-K=0.5-1.5 region, then we make the assumption than these sources with J-K_s> 2 are background stars. The extinction of Cham I is in the range between 1 and 8 (peak at 10) (Cambrésy, this issue), then, sources between K_s=8 and K_s= 13 and (J-K_s)=1-2 could be T Tauri stars without IR excess while sources with a (J-K)>2 and brighter than K_s=13 could be T-Tauri with a strong IR excess.

2. Discussion

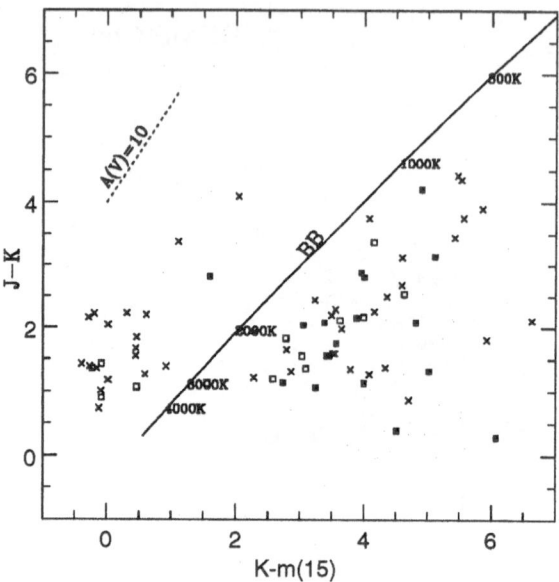

Figure 3. (J-K), (K-LW2) diagram the common sources detected both by DENIS and ISOCAM. The positions of the known CTT (filled square) and WTT(open square) are reported. This diagram separates two distinct groups of sources (see text)

2.1. COLOR-COLOR DIAGRAMS

The J-K vs K-m(15) diagram presented in Fig. 3, shows two distinct groups of sources: *i*) K-m(15)=0 (only WTT fill this part of the diagram); *ii*) K-m(15)>2. indicate an IR excess at 15 μmfor these sources (presence of a dust circumstellar disk ?). CTT and WTT are located in this region of the diagram. This suggests that WTT can be either Class II sources with IR excess, or Class III sources. The same conclusion is drawn out using J-K vs m(6.75)-m(15) diagram.

2.2. SPECTRAL ENERGY DISTRIBUTION

Using the two instruments we can derive a spectral energy distribution (SED) on a large spectral domain. The SED of the known CTTs and WWTs detected by DENIS and ISOCAM are presented Fig. 4. We remark 3 differents groups: CCTs which are class II sources with a spectral index n ($0 < n < 1$), WTTs with and without IR excess.

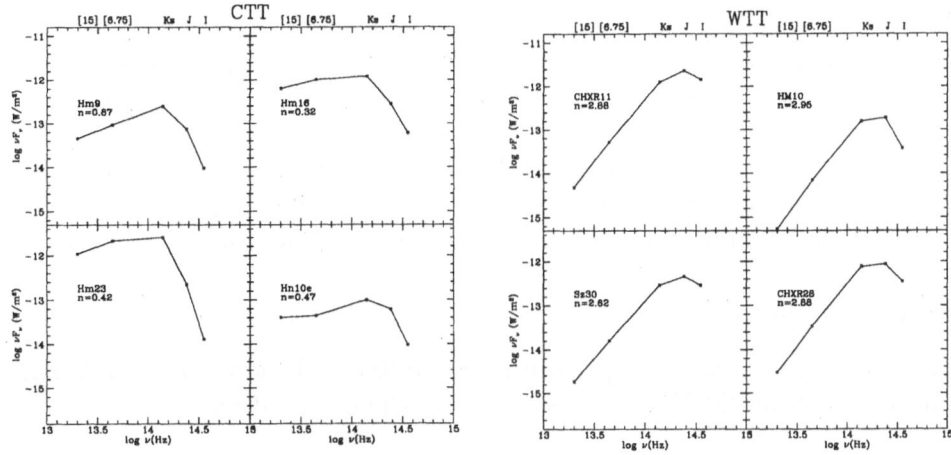

Figure 4.   Left: Energy distribution of known CTTs. right: SED of WTTs with infrared excess.

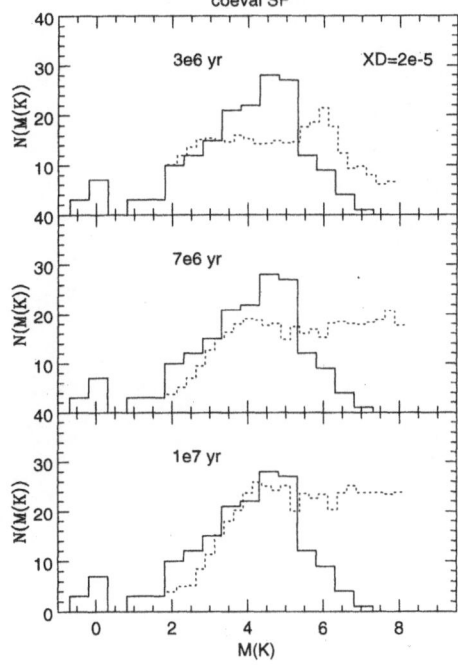

Figure 5.   Observed LF in K of the ISOCAM-DENIS sources (solid line), the dashed line shows the model for a coeval SF modele computed for a deuterium abundance XD=2e-5

2.3. AGE OF THE CHAM I YSO

We have derived a K luminosity function for coeval star formation model (shown in Fig. 5) and continous SF models from the stellar evolution tracks computed for different deuterium abundances (D'Antona, private communication), and taking a distance d=140pc for the cloud and a mean A_V = 4. From the comparaison of these different models, it appears that the young stellar population has an age not younger that 10^7 year.

Acknowledgements

We thank F. D'Antona for providing the stellar evolution models in K band, and the La Silla DENIS team for making the observations.

References

Cambrésy L., Epchtein N., Copet E., De Batz B., Kimeswenger S., Le Bertre T., Rouan D. and Tiphène D. (1997) *A&A*, **324**, p.L5
Cesarsky C (+ 20 authors) (1996) *A&A*, **315L**, p. 32
Nordh L (+20 authors) (1996) *A&A*, **315L**, p. 185

PLANETARY NEBULAE WITH DENIS

S. KIMESWENGER

Institut für Astronomie der Leopold–Franzens Universität
Technikerstraße 25, A–6020 Innsbruck, AUSTRIA

1. Introduction

The last, and only, attempt to survey the sky in the near infrared range was the TMSS (Neugebauer & Leighton, 1969). This survey contains mainly bright stars. The Deep Near Infrared Southern Sky Survey (DENIS) is the first attempt to survey all the southern sky in the near infrared (NIR) range in three bands; I, J and Ks (Epchtein et al., 1994). Planetary nebulae (PNe) were investigated by means of aperture photometers (e.g. Whitelock 1985, Kwok et al. 1986, Pena & Torres–Peimbert 1987, Preite–Martinez & Persi 1989, Phillips & Cuesta 1994) in the past. These investigations often use J, H and K bands. Thus the DENIS survey will lead, due to different bands and the total sky coverage, to a new view on PNe in this wavelength domain. We show here the capabilities of investigations of PNe with the DENIS data, being comparable in spatial resolution, will also support investigation at longer wavelengths (Kimeswenger et al., 1997a) done with the ISOCAM instrument (Cesarsky et al., 1996). Spatially resolved observations also provide better information about the contamination of the red (or highly reddened) foreground stars. The survey also will uncover the nature of several objects suspected to be PNe by means of their IRAS colors, but having no optical identification yet.

2. Individual objects

The objects here were selected from the pool of already observed PNe with diameters greater than 15 arcseconds. A set of them have NIR information in the literature already. The objects here were selected to show the main advantages of NIR imaging.

171

N. Epchtein (ed.),
The Impact of Near-Infrared Sky Surveys on Galactic and Extragalactic Astronomy, 171-174.
© 1998 *Kluwer Academic Publishers.*

2.1. NGC 2440 (= PN G234.8+02.4)

This object is given with $m_J = 10^m33$ and $m_K = 9^m68$ (Whitelock, 1985) and is one of the classical bipolar systems (sometimes even classified as quadrupolar). The nebula has 16" in diameter (Acker et al. (1992)). The DENIS images of NGC 2440 (Fig. 1) show the main nebula being already significantly larger (22" × 28").

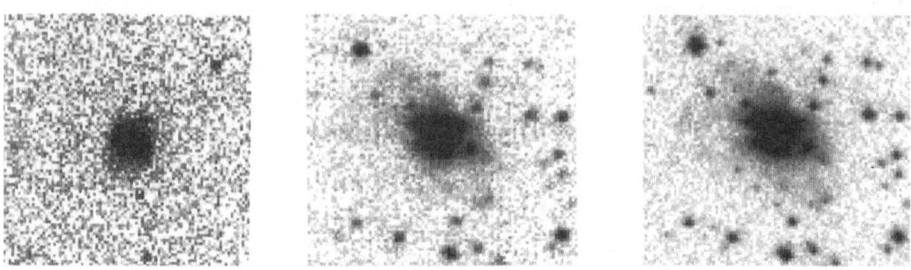

Figure 1. Images of NGC 2440 by the DENIS survey at the ESO 1m (from left to right: Ks, J, I) with a resolution of 1"/pixel and an exposure time of 10 seconds.

The photometrical result obtained with DENIS in the K band is somewhat fainter than the values listed in the literature. The J band corresponds extremely well. The total I band flux had to be slightly corrected for stellar contamination (0^m08). To avoid confusion due to different aperture sizes, we obtained not only the total flux, but also that one of the apertures used in the photometers in other studies.

TABLE 1. Results of the photometry for NGC 2440 obtained with DENIS and found in the literature

band	aperture ["]	DENIS [mag]	from lit. [mag]	ref.
K/Ks	total	9,38		
	27	9,52	9,39	Persson & Frogel (1987)
	24	9,63	9,68	Whitelock (1985)
	14	10,22	10,07	Pena & Torres–Peimbert (1987)
J	total	9,99		
	24	10,21	10,33	Whitelock (1985)
	14	10,71	10,67	Pena & Torres–Peimbert (1987)
I	total	11,21		

2.2. NGC 3918 (= PN G294.6+04.7)

This system is a "classical" round system without any structure at a scale of a few arcseconds. The ring structure is hardly visible. It is more likely a uniform brightness object. Persi et al. (1987) have given $m_K = 8\overset{m}{.}85$ and $m_J = 9\overset{m}{.}13$ for this object. The aperture used there is not clear to us, but taking into account the size of the object of 19" only, we assume that the whole object was within the aperture. We find $8\overset{m}{.}71$, $9\overset{m}{.}18$ and $10\overset{m}{.}58$ for K, J and I respectively.

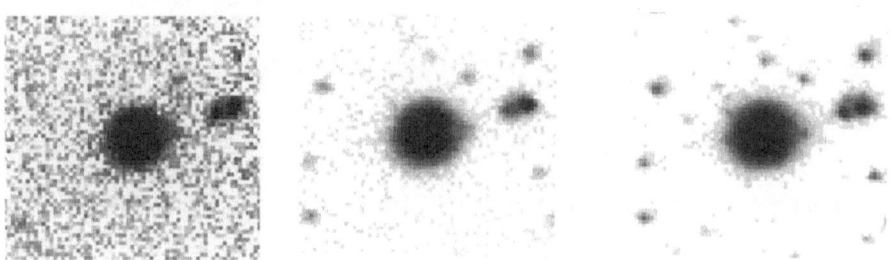

Figure 2. Images of NGC 3918 by the DENIS survey (from left to right: Ks, J, I).

2.3. KFL 14 (= PN G002.5-05.4)

This nebula is claimed to be a bulge sample object (Acker et al., 1992). The membership investigations of the bulge sample are of importance for the calibration of many "global" properties of PNe. The extinction towards the nebulae is known (by Balmer decrements). It is found to be less than $E_{B-V}=1\overset{m}{.}$. Comparing the red DENIS images with the (deeper !) sky survey clearly shows, that the extinction towards the bulge is significantly higher than $E_{B-V}=1\overset{m}{.}$. Thus, this kind of investigations of the "surrounding" of PNe gives us a new tool to better select the real bulge sample.

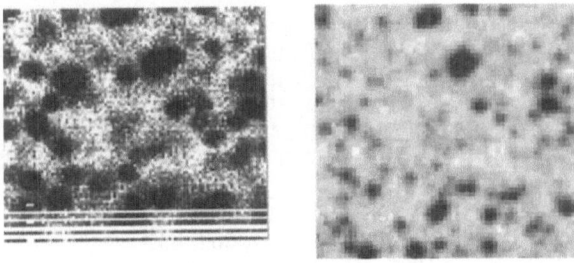

Figure 3. The images of KFL 14 (left: DENIS J, right: optical sky survey).

2.4. SAKURAI'S OBJECT (= PN G010.4+04.4)

This object, first claimed by Y.Sakurai in Feb. 1996 to be a nova, is a born–again PNe undergoing a very late He-flash. The DENIS images allowed (together with spectra from the LasCampanas 100" telescope) to monitor the dust formation phases (Kimeswenger et al., 1997b).

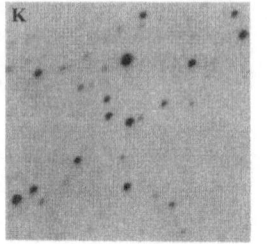 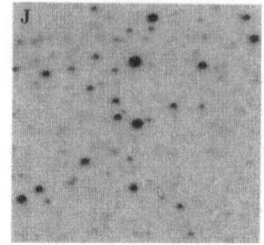 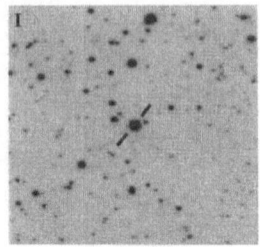

Figure 4. The images of Sakurai's Object obtained in spring 1996.

Acknowledgements
This project was supported by the FWF projects P8700-PHY and P10036-AST. The DENIS project is partly funded by the European Commision through *SCIENCE and Human Capital and Mobility* plan grants. It is also supported, in France by the Institut National des Sciences de l'Univers, the Education Ministery and the Centre National de la Recherche Scientifique, in Germany by the State of Baden–Württemberg, in Spain by the DGI-CYT, in Italy by the Consiglio Nazionale delle Ricerche, in Austria by the *Fonds zur Förderung der wissenschaftlichen Forschung und Bundesministerium für Wissenschaft und Forschung*, in Brazil by the Fundation for the development of Scientific Research of the State of São Paulo (FAPESP), and in Hungary by an OTKA grant and an ESO C & EE grant.

References

Acker, A., Ochsenbein, F., Stenholm, B., et al: 1992, The Strasbourg-ESO Catalogue of Galactic Planetary Nebulae, ESO, Munich, Germany
Cesarsky C.J., Abergel A., Agnèse P., et al., 1996, A&A, 315, L32
Epchtein, N., de Batz, B., Copet, E., et al., 1994, Ap&SS 217, 3
Neugebauer G., Leighton R.B., 1969, Two Micron Sky Survey, NASA, SP 3047
Kimeswenger, S., Kerber, F., Gratl, H., et al.: 1997, IAUC 6608
Kimeswenger, S., Kerber, F., Weinberger, R.: 1997, MNRAS, submitted
Kwok, S., Hrivnak, B.J., Milone, E.F.:1986, ApJ, 303, 451
Pena, M., Torres–Peimbert, S.: 1987, RMexAA, 14, 534
Persi, P., Preite–Martinez, A., Ferrari–Toniolo, Spinoglio, L.: 1987, ApSS Library, 135, 221
Persson, E.S., Frogel, J.A.: 1973, ApJ, 182, 503
Phillips, J.P., Cuesta, L.: 1994, A&AS, 104, 169
Preite–Martinez, A., Persi, P.: 1989, A&A, 218, 264
Whitelock, P.A.: 1985, MNRAS, 213, 59

V- Extragalactic

GALAXIES WITH DENIS:

Preliminary star/galaxy separation and first results

GARY A. MAMON
Institut d'Astrophysique de Paris & DAEC, Obs. de Paris
98 bis Bd Arago, F-75014, Paris, FRANCE

AND

JEAN BORSENBERGER, M. TRICOTTET AND V. BANCHET
Institut d'Astrophysique de Paris

Abstract. The numerous extragalactic and cosmological motivations of the DENIS and 2MASS near infrared surveys are outlined. The performance of the DENIS survey is estimated from $50\,\mathrm{deg}^2$ of high galactic latitude data $(20° < |b| < 60°)$. Simple star/galaxy separation methods are presented and comparison with 300 visually classified objects as well as COSMOS and APM classifications. We find that the peak intensity over isophotal area is an excellent star/galaxy separation algorithm, fairly robust to variations of the PSF within the frames, achieving 98.5% completeness and 92.5% reliability for $I < 16.5$, in comparison with visual classification. A new estimate of the photometric accuracy for galaxies is presented. The limiting factors for homogeneous galaxy extraction at high galactic latitudes are completeness and photometric accuracy in K, photometric accuracy in J and star/galaxy separation in I (also used for classification in J and K). Galaxy counts are presented on $50\,\mathrm{deg}^2$. The I counts are in excellent agreement with a Euclidean extrapolation of the published counts around $I = 16 - 17$ (more so than in all previous studies), and thus point to a high normalization at the bright end, in contrast with the counts published from the APM and COSMOS plate scans. The J-band differential galaxy counts follow the relation $N(J) = 12 \pm 1\,\mathrm{dex}(0.6\,[J - 14])\,\mathrm{deg}^{-2}\mathrm{mag}^{-1}$. Extrapolation of these high latitude counts suggest that DENIS will produce highly homogeneous catalogs of $\simeq 6000$ $(K < 11)$, $\simeq 700\,000$ $(J < 14.8)$ and, $\simeq 1\,000\,000$ $(I < 16.5)$ galaxies, respectively with photometric accuracy of 0.08^m in I and 0.20^m in J and K. Larger highly homogeneous samples are expected with improvements to the camera and the algorithms.

N. Epchtein (ed.),
The Impact of Near-Infrared Sky Surveys on Galactic and Extragalactic Astronomy, 177-192.
© *1998 Kluwer Academic Publishers.*

1. Introduction

The DENIS consortium has been imaging the southern sky in the I (0.8μm), J (1.25μm) and K_s (2.15μm) wavebands since December 1995. When the survey is complete, around 2000–2001, we expect to have extracted tens of thousands of galaxies in K, roughly one million in J, and a few million in I (see § 7 below for our estimated sizes of homogeneous, highly complete, reliable and photometrically accurate galaxy catalogs).

Much of the information in this review has been given elsewhere (Mamon et al. 1997b). The notable improvements here are improved reliability estimates from a much larger visually classified sample, a first-order optimization of star/galaxy separation yielding a one-half magnitude improvement in the high completeness/reliability magnitude limit and a more accurate estimate of the photometric accuracy.

2. Prospective scientific impact

Wide-angle near infrared (hereafter NIR) galaxy surveys, such as DENIS and 2MASS (see Schneider, Jarrett, Rosenberg and Cutri, all in these proceedings) will have a wide array of scientific prospects, of which a few are listed below. The two important advantages of NIR selection are 1) the near transparency of interstellar dust in our foreground Galaxy and within external galaxies, and 2) the low sensitivity of NIR light to recent star formation in galaxies (see Mamon et al. 1997b), hence a better estimation of the stellar mass content of galaxies in the NIR.

Statistics of NIR properties of galaxies: DENIS and 2MASS will provide the first very large galaxy databases with NIR photometry. Photometry of the brighter galaxies will be coupled with redshift measurements, either already made, or performed during spectroscopic followups (see, e.g., Mamon 1996; Paturel, in these proceedings) to be used for distance estimates and computation of precise parameters of the fundamental plane and Tully-Fisher relations (see Vauglin et al. 1997; Rosenberg, in these proceedings).

Cross-identification with other wavelengths: The extragalactic objects extracted by DENIS and 2MASS will be cross-identified with analogous samples at other wavelengths, such as optical galaxy samples, for example in the Zone of Avoidance (see Kraan-Korteweg et al., in these proceedings), IRAS galaxies (Saunders et al. 1997), quasars (see Cutri, in these proceedings), radio-galaxies, galaxies found in blind HI surveys (see Kraan-Korteweg et al., in these proceedings), etc. The NIR properties (mainly their location in color-color diagrams) of such objects will be targeted for discovering new large samples of such objects. One should expect followups at non-NIR wavelengths of DENIS and 2MASS galaxies.

Galaxy counts: There has been a debate on the level of galaxy counts at the bright end, as first estimates (Heydon-Dumbleton et al. 1989; Maddox et al. 1990) found a depletion relative to the extrapolation of the faint-end counts, while later work (e.g. Bertin & Dennefeld 1997) disputed this. This debate has consequences on galaxy evolution and on whether the environment of the Local Group is underdense on very large scales ($z \lesssim 0.1$).

Zone of avoidance   There are two main applications for studying galaxies behind the Galactic Plane (see Kraan-Korteweg et al., in these proceedings): 1) Mapping the large-scale distribution of galaxies in this still poorly known region. Indeed, the Zone of Avoidance contains interesting structures such as the largest large-scale concentration of matter in the local Universe, the Great Attractor (at the intersection of the Supergalactic Plane and the Galactic Plane, Kolatt, Dekel & Lahav 1995) and within the Great Attractor, the Norma cluster, Abell 3627, richer and closer than the Coma cluster (Kraan-Korteweg et al. 1995). 2) The fluxes and angular sizes of galaxies are affected by extinction from dust in the Galactic Plane, and one can measure this extinction from galaxy counts (Burstein & Heiles 1982), colors (Mamon et al. 1997a), and color-color diagrams (Schröder et al. 1997, and Kraan-Korteweg et al., in these proceedings).

Small-scale structures of galaxies   Only a few catalogs of clusters (Lumsden et al. 1992; Dalton et al. 1997; Escalera & MacGillivray 1995, 1996) and compact groups (Prandoni, Iovino & MacGillivray 1994) are based upon automatically selected galaxy samples, which happen to be optical and photographic (hence subject to photometric non-linearities). Because star formation is probably enhanced by galaxy interactions, one expects that the statistical properties of pairs, groups and clusters of galaxies built from NIR selected galaxy catalogs will be different from those built from optical catalogs. DENIS and 2MASS will thus have the double advantage of using a NIR galaxy sampled based upon linear (non-photographic) photometry. The applications of such NIR-based samples of structures of galaxies are numerous (e.g. Mamon 1994) and include understanding the dynamics of these structures, their bias to projection effects, their constraints on Ω_0 and the primordial density fluctuation spectrum, their use as distance indicators, and the environmental influences on galaxies.

Large-scale structure of the Universe: The NIR selection and the linear photometry will also benefit the measurement of statistics (two-point and higher-order angular correlation functions, counts in cells, topological genus, etc.) of the large-scale distribution of galaxies in the Universe. For example, the (3D) primordial density fluctuation spectrum of galaxy clustering can be obtained from the two-point angular correlation function (Baugh & Efstathiou 1993) or from the 2D power spectrum (Baugh & Efstathiou 1994). Moreover, by the end of DENIS and 2MASS, large-scale cosmological simulations with gas dynamics incorporated (thanks to which galaxies are properly identified) will provide adequate galaxy statistics in projection that will be compared with those obtained from the surveys, iterating over the cosmological input parameters of the simulations.

3. Galaxy extraction and current galaxy pipeline

The current galaxy pipeline consists of the following steps:

1) Bias subtraction, flat-fielding, bad pixel mapping and astrometric calibration (standard DENIS Paris Data Analysis Center pipeline, Borsenberger 1997); 2) Cosmic ray removal; 3) Extraction of photometric zeropoints and airmasses from relevant files; 4) Galaxy extraction using the *SExtractor* (Bertin & Arnouts 1996) object extraction software, version 1.2b6a (which includes a neural-network star/galaxy separator, Bertin 1996, whose input parameters are 8 isophotal areas, the maximum intensity and as a control parameter, the FWHM of the PSF), with detection and Kron (1980) photometry parameters optimized from simulated images.

4. Star/galaxy separation

Nevertheless, star/galaxy separation is intrinsically difficult because, at the galaxy extraction limits $I \simeq 16.5$ (see below), DENIS will extract roughly 5.5 times as many stars as galaxies in I, at very high galactic latitude ($|b| \simeq 70°$, see Lidman & Peterson 1996), and the ratio worsens considerably at lower galactic latitudes and at brighter magnitudes.

We discuss below the steps towards an efficient star/galaxy separation method. For this, we extracted in the I band (which has the best angular resolution) classical star/galaxy separation diagnostics such as isophotal area, peak intensity, and FWHM, as well as the neural-network based stellarity parameter, in a direct fashion, or using a suitably modified version of SExtractor that includes a two-dimensional modeling of the PSF that is used as input to the neural network.

Figure 1 shows how these quantities vary with magnitude for all objects at least 20 pixels from the frame edges on a high latitude strip.

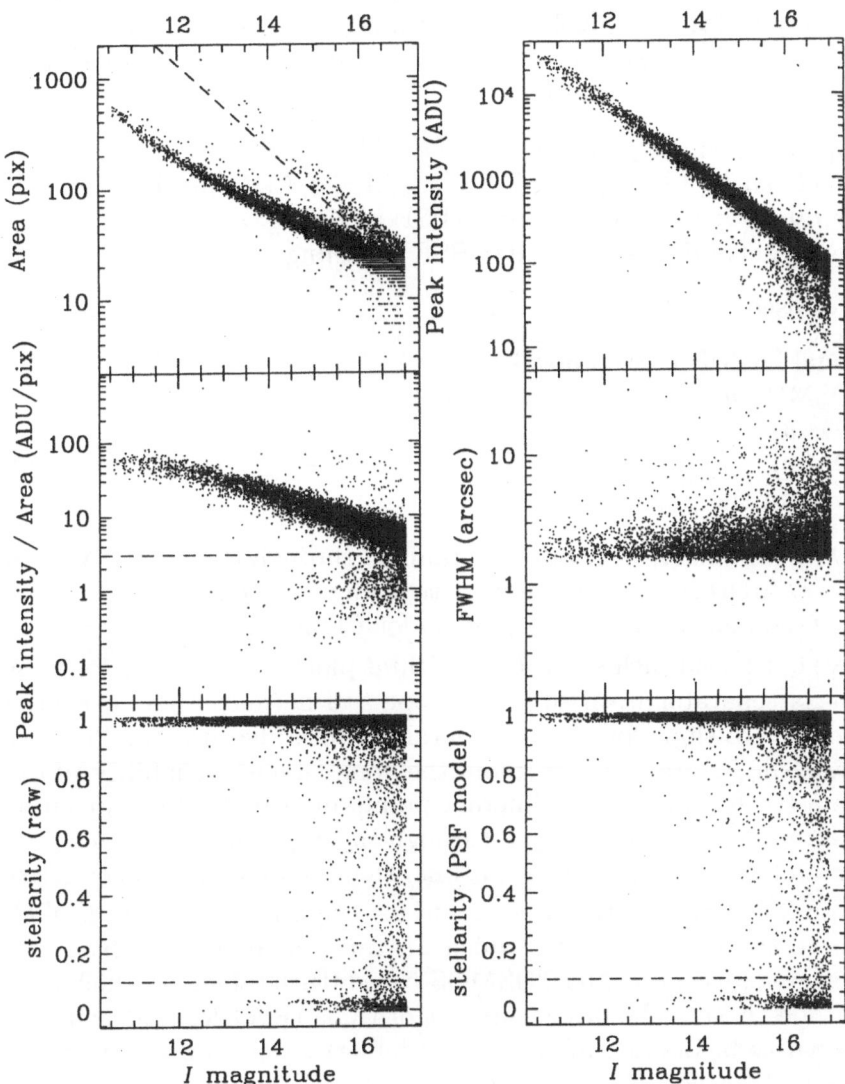

Figure 1. Diagnostics of star/galaxy separation over one DENIS ($6\,\mathrm{deg}^2$) strip. The *dashed lines* are the critical lines for selection of candidates for visual classification (see § 4.1).

4.1. TESTS OF AUTOMATIC STAR/GALAXY SEPARATION

One of us (G.A.M.) has classified by eye a set of 329 galaxy candidates on 109 DENIS *I* band images (of which 33 appeared on consecutive images, leaving 296 unique candidates). These candidates were chosen with $I \leq$ 16.5, centers at least 20 pixels from the image borders. Furthermore, they met *at least one* of the following loose (to ensure completeness) galaxy criteria (*dashed lines* in Fig. 1):

- Isophotal area: $A \geq 40 \, \mathrm{dex}[-0.38(I - 16)]$ pixels
- Pseudo surface brightness: $\Sigma = I_{\mathrm{peak}}/A \leq 3$ ADU/pixels
- Neural-network stellarity before PSF modeling: $s_0 \leq 0.1$
- Neural-network stellarity after PSF modeling: $s \leq 0.1$

We've used 5 sets of truth tables:

- Visual DENIS I (see above)
- COSMOS b_J
- APM b_J
- APM r_F
- A mix of the previous 4

The COSMOS and APM lists were obtained through the World Wide Web (`telnet://catalogues apm3.ast.cam.ac.uk` for the APM and `telnet://cosmos cosmos.aao.gov.au` for COSMOS).

We've optimized each of the 6 algorithms plotted in Figure 1 for a linear star/galaxy separator in these plots (slope and normalization, except that we forced a zero slope for the two neural network algorithms). The results are showed in Figure 2, which plots the completeness-reliability plots for 4 of the 5 truth tables. The different points in Figure 2 for a given algorithm correspond to different cuts through the algorithm versus magnitude diagram and we only plotted the optimal slope, varying the normalization.

Figure 2 shows that the pseudo-surface brightness criterion is slightly superior to the peak intensity, which, in turn, is slightly superior to the isophotal area (except for the COSMOS-based truth table, for which isophotal area does best). The other three algorithms (FWHM, and neural network stellarity before and after PSF modeling), are far inferior to the first three algorithms. For the visually classified DENIS I sample, we achieve 92.5% reliability at 98% completeness, and for the global sample we obtain 96% reliability at 96% completeness. The poor results of the neural networks is probably due to the variations of the PSF across the frames, and for this particular DENIS strip (number 5570), PSF modeling worsened the results!

4.2. COSMOS AND APM VERSUS VISUAL STAR/GALAXY SEPARATION

Table 1 shows the comparison between the visual classification and the classification obtained from the COSMOS and APM lists.

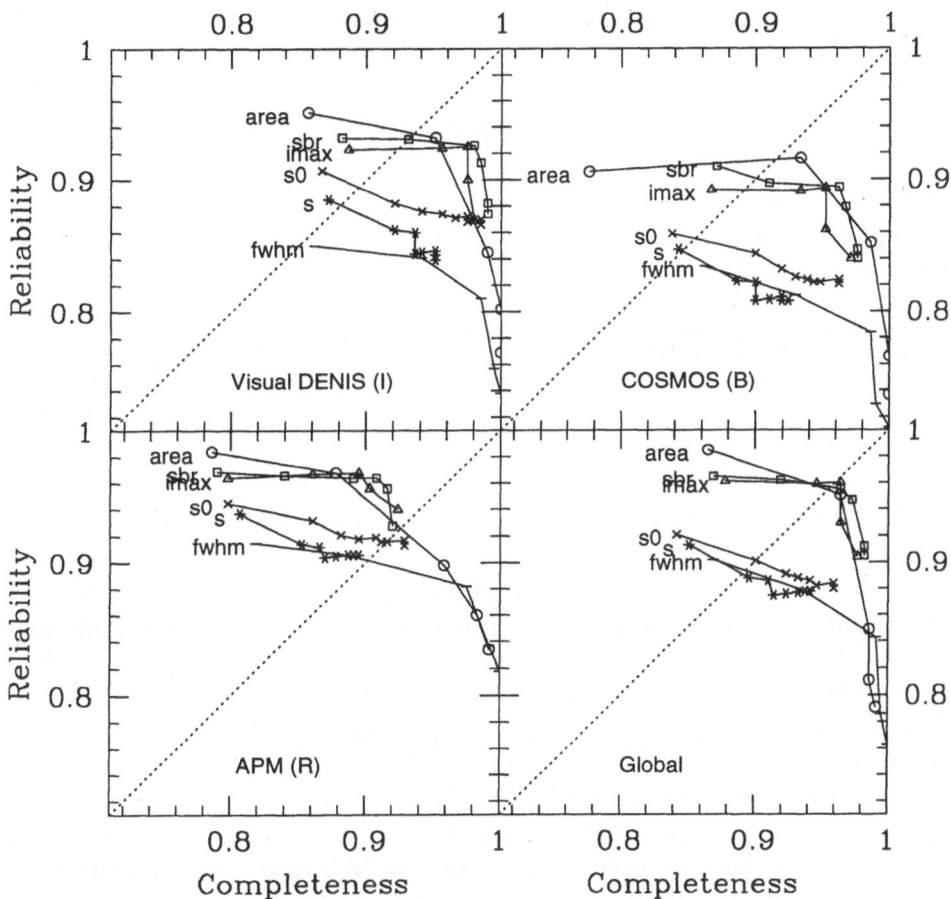

Figure 2. Completeness versus reliability of different automatic star/galaxy separation algorithms using 4 different truth tables. The algorithms are isophotal area (`area`), peak intensity (`imax`), pseudo-surface brightness (`sbr`), star/galaxy separation without PSF modeling (`s0`), star/galaxy separation with PSF modeling (`s`), and full-width half maximum (`fwhm`).

Of the 11 objects termed as junk, 3 were fragments of a bright galaxy, two were deemed optical flaws, but according to both APM and COSMOS, one of those was a star.

The numbers in Table 1 do not permit to establish which star/galaxy separation is best between visual DENIS, APM or COSMOS. However, if one assumes that visual DENIS star/galaxy separation is perfect, one would then conclude that APM and COSMOS both have a completeness of $193/203 = 95\%$ at $I = 16.5$ (this also assumes that the DENIS I extraction

TABLE 1. Visual DENIS I versus COSMOS and APM star/galaxy separation

Visual DENIS I Type	Total	COSMOS b_J Galaxy	Star	Notfound	APM b_J Galaxy	Star	Faint	Notfound
Galaxy	203	193	10	0	193	5	3	2
Star	53	6	46	1	9	42	1	1
Star+Star	8	4	3	1	4	3	0	1
Faint	21	7	11	3	10	10	0	1
Junk	11	0	2	9	2	3	3	3
Total	296	210	72	14	218	63	7	8

is 100% complete, which remains to be proven). The reliability of the extraction would then be $193/210 = 91\%$ for COSMOS and $19?/218 = 89\%$ for APM.

If one assumes that APM or COSMOS are complete, than the incompleteness of the DENIS galaxy extraction can be estimated from the objects too faint for DENIS visual classification but called galaxies by the optical surveys. One obtains completeness levels of 95% or 97% at $I = 16.5$ using APM or COSMOS, respectively. Of course, if the visual classification were imperfect and that objects classified as stars or double stars are in fact galaxies, the completeness of DENIS visual classification would decrease to levels of 90% or 92% using APM or COSMOS, respectively. Moreover, DENIS may not have detected objects at $I = 16.5$ that are seen in the optical surveys, and this issue will be addressed in a forthcoming publication.

4.3. QUICK AND DIRTY AUTOMATIC STAR/GALAXY SEPARATION

Since the pseudo-surface brightness criterion seems to produce the best star/galaxy separation, we have adopted the following preliminary algorithm for each DENIS strip:

We adopt a constant critical pseudo surface brightness (independent of magnitude — the optimal slope with respect to the visual DENIS I, COSMOS B, APM B, and global classifications was 0.05), by fitting with a cubic polynomial the histogram of the values of $\Sigma = I_{\text{peak}}/A$ for $I \leq 16.5$, in a range chosen to exclude the peak due to the stars. Although $k\sigma$ curves down from the stellar locus have negative slope, the higher slope of the galaxy counts relative to the star counts leads us to believe that a given reliability will be achieved with a cut of k that decreases with magnitude, $i.e.$, with a lower slope for Σ_{crit}. This may explain why the optimal slope is non-negative.

For the J and K bands, we rely on the star/galaxy separation performed in the I band. Because the I band has better angular resolution and is more sensitive than J or K (except at very low galactic latitudes, corresponding to visual extinction $A_B > 3$, see Kraan-Korteweg et al., in these proceedings), using I-band star/galaxy separation is superior to doing star/galaxy separation directly in J or in K.

Our star/galaxy separation, relying only on pseudo surface brightness is simpler than in our previous work (Mamon et al. 1997b), where we required out galaxies to satisfy both neural network stellarity (after PSF modeling) and isophotal area algorithms, and our former star/galaxy separation method had the disadvantage of using a fixed critical isophotal area line, whereas strip to strip variations of the PSF lead to variations of this critical line from one strip to another.

We have thus analyzed a little over $50\,\mathrm{deg}^2$ of DENIS data, restricting ourselves here to $I < 17$.

5. Photometry

We estimate below the accuracy of DENIS galaxy photometry using objects within image overlaps and comparing with APM and COSMOS, and we use color-magnitude diagnostics as an additional test on the reliability of star/galaxy separation.

5.1. PHOTOMETRIC ACCURACY FROM OVERLAPS

Figure 3 shows the magnitude differences on unflagged overlap objects extracted from $50\ \mathrm{deg}^2$ of high galactic latitude data.

Contrary to the analogous figure in Mamon et al. (1997b), we have high certainty on the extragalactic nature of the J-band and K-band overlap objects (since again, we rely on I-band star/galaxy separation). For this reason, the photometric accuracy is worse than given in Mamon et al. (1997b): The rms error on a single measure is 0.05 at $I = 15$, 0.10 at $I = 17$, 0.10 at $J = 13.7$, and 0.20 at $J = 14.8$. There are too few K overlaps to conclude strongly, but indications (based upon only 4 points!) are that the rms photometric accuracy for a single measure is roughly 0.20 at $K \simeq 12.2$. The J-band photometric accuracy was considerably better in our previous study (Mamon et al. 1997b), but unreliable direct (using neural network stellarity in J lower than 0.2) star/galaxy separation had been used for the photometric accuracy study of that work, and the inclusion of stars tends to improve the photometric accuracy.

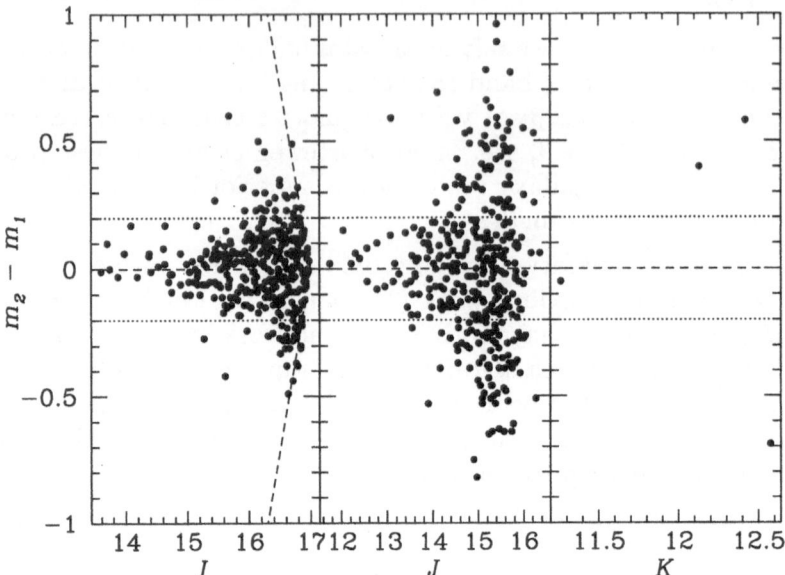

Figure 3. Photometric accuracy for galaxies lying within overlaps of 2 images, extracted within 50 deg² of high galactic latitude (20° < |b|60°) DENIS data. Objects closer than 20 pixels to the frame edges are excluded). The *tilted dashed line* represents an $I \leq 17$ selection, whose effects are also seen in the J band.

5.2. COMPARISON OF DENIS GALAXY PHOTOMETRY WITH COSMOS

For the 3.6 deg² region in which we visually classified our extracted objects, we plot in Figure 4 the color-magnitude relation obtained with COSMOS b_J photometry, taken from the World Wide Web. This figure shows the difficulties in star/galaxy separation, as a number of points lie far off the $B - I \simeq 2 - 3$ region. Part of this difficulty lies in poor star/galaxy separation from COSMOS. Moreover, there is a trend for bluer galaxy colors at brighter magnitudes, which we interpret as poor photometry on the COSMOS side, because of inaccurate compensation for plate saturation.

We also attempted the same with APM data from the World Wide Web, but that photometry suffers from unusually strong systematic errors at the bright end (up to 6 mag difference with COSMOS!), as the photometric calibration has been optimized for stars that saturate at these magnitudes (Maddox, private communication).

5.3. COLORS OF DENIS GALAXIES

Figure 5 shows the color-magnitude diagram for the galaxies. The bluest two points turn out to be galaxies! Visual inspection shows that they are

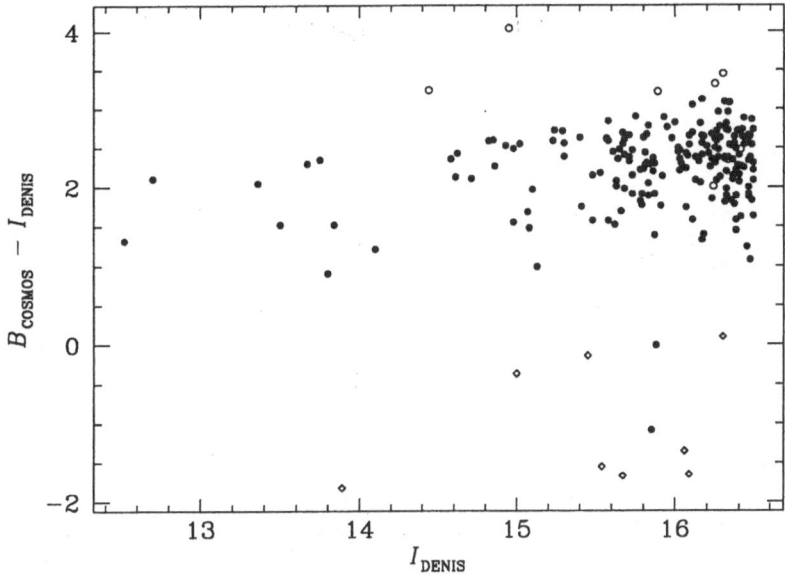

Figure 4. Color-magnitude diagram for galaxies extracted within 3.6 deg^2 of high galactic latitude ($20° < |b|60°$) DENIS data. *Diamonds* refer to objects classified as stars by COSMOS, but as galaxies by APM, DENIS visual inspection and all DENIS automatic star/galaxy separation algorithms. *Open circles* are objects classified as galaxies with low certainty by the DENIS visual classification and that were not stars in COSMOS.

low surface brightness galaxies that are barely visible in J (and invisible in K). The use of adaptive aperture photometry to define colors makes such objects appear very blue. We checked that their central colors are normal.

Figure 5 shows that at the limit $J = 14.8$ for $\Delta J = 0.20$ mag photometric accuracy, the star/galaxy separation performed in I should be roughly as reliable as at $I = 16.5$, and could be made even more reliable by culling out the reddest objects for which $I > 16.5$.

In Figure 6, we plot the color-color diagram for extracted galaxies. The galaxy colors cluster around $I - J = 1.2 \pm 0.3$, $J - K = 1.1 \pm 0.5$, but there are indications for fairly bright objects with red $J - K \simeq 2$ colors, which upon visual inspection are confirmed as galaxies. An important fraction of the points off the central cluster lie near the frame edges where the PSF is larger. The large open circle refers to an object too faint in I for reliable star/galaxy separation, and indeed, visual inspection shows it to be a star blended with a faint galaxy.

6. Galaxy counts

Figure 7 illustrates our IJK galaxy counts. The K-band counts become

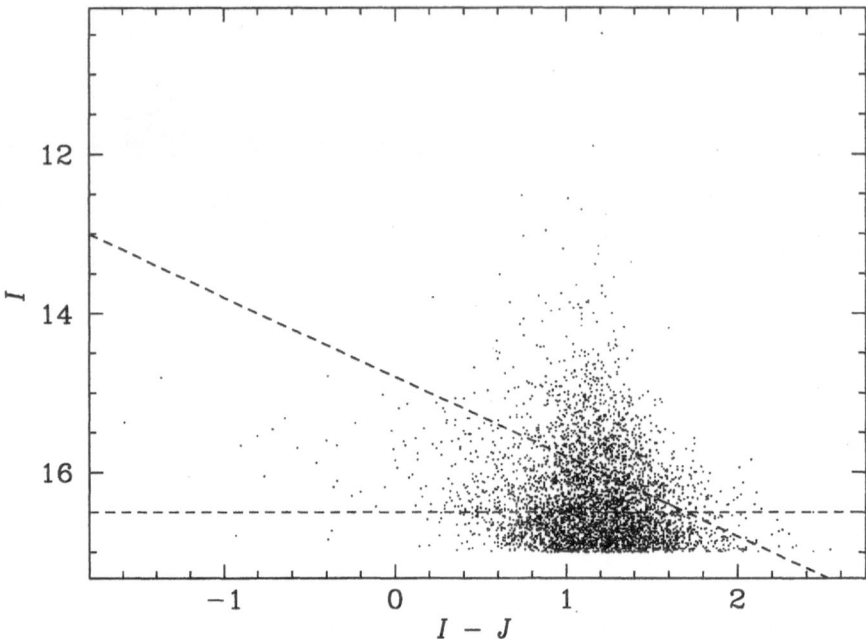

Figure 5. Color-magnitude diagram for galaxies extracted within 50 deg^2 of high galactic latitude ($20° < |b| < 60°$) DENIS data. Objects closer than 20 pixels to the frame edges are excluded. The *horizontal line* represents $I = 16.5$ (the and the dashed line represents $J = 14.8$ (the limit for 0.20 mag J-band photometry and reliable star/galaxy separation).

incomplete at $K \simeq 11$, in comparison to both published counts by Gardner et al. (1996) and to the expected Euclidean 0.6 slope (the completeness is still roughly 50% at $K = 12$).

The I band counts match well the published data, although Lidman and Peterson (1996) find fewer counts at the bright end, while Gardner et al. find more counts at the bright end (the two sets of published data differ by a factor of 3 at $I < 15$). Note that DENIS, Gardner *et al.* and Lidman & Peterson all work with the Cousins I band, so no conversion was made from another I filter. Also, our survey has smaller error bars at the bright end as it covers 4 to 5 times the solid angle of the two cited surveys. Our bright-end I-band counts are more consistent with the extrapolation of the faint counts with a Euclidean slope than either two sets of published data (our high value at $I = 16.5$ is caused by important stellar contamination in the fainter half of the bin; also, at $I > 18$, the published counts become lower than the Euclidean line because of significant k-correction at these magnitudes). In this sense, although not as high as Gardner et al.'s counts, *the DENIS I-band counts argue for a high bright-end normalization, consistent with little*

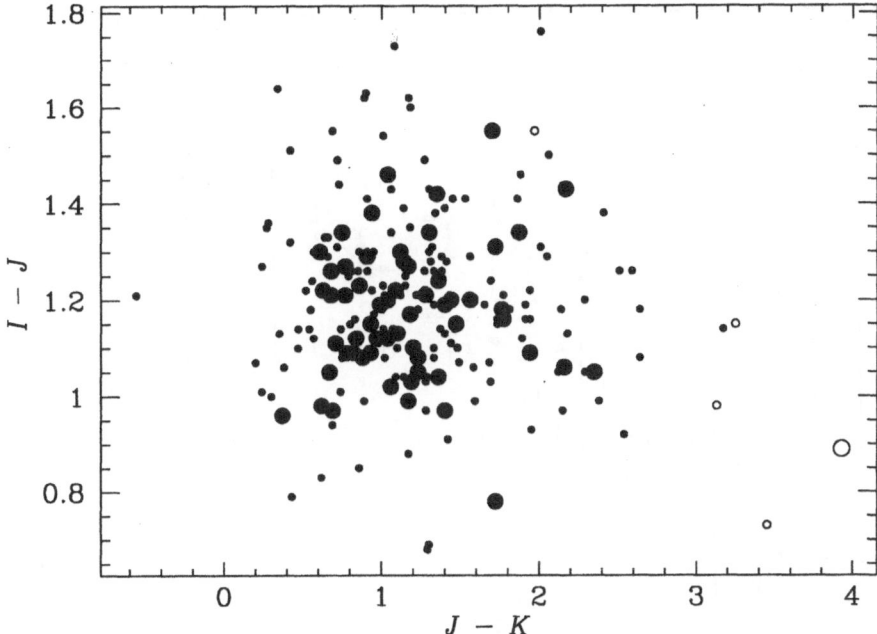

Figure 6. Color-magnitude diagram for galaxies extracted within 50 deg^2 of high galactic latitude ($20° < |b| < 60°$) DENIS data. Objects closer than 20 pixels to the frame edges are excluded. *Large* and *small circles* are for objects brighter or fainter than $K = 12$ (the limit for fairly accurate K photometry, see § 5.1), respectively. *Filled* and *open circles* correspond to objects brighter and fainter than $I = 16.5$ (the rough limit for reliable star/galaxy separation, see § 4.1), respectively.

galaxy evolution at the bright end, in line with analogous findings by Bertin and Dennefeld (1997) using blue counts.

The J counts are new (although they were already shown in Mamon et al. 1997b). They are highly complete to $J = 15$, follow very well the Euclidean slope of 0.6, and are well described by the relation $N(J) \simeq 12 \times \text{dex}[0.6\,(J - 14)]\,\text{deg}^{-2}\,\text{mag}^{-1}$.

7. Discussion

From the results of the preceding sections, we can establish limits for the homogeneous extraction of galaxies from DENIS, as given in Table 2.

The limiting factors turn out to be star/galaxy separation in I, photometry and star/galaxy separation in J, and detection in K (assuming that I-band star/galaxy separation is used to classify objects detected in the other bands).

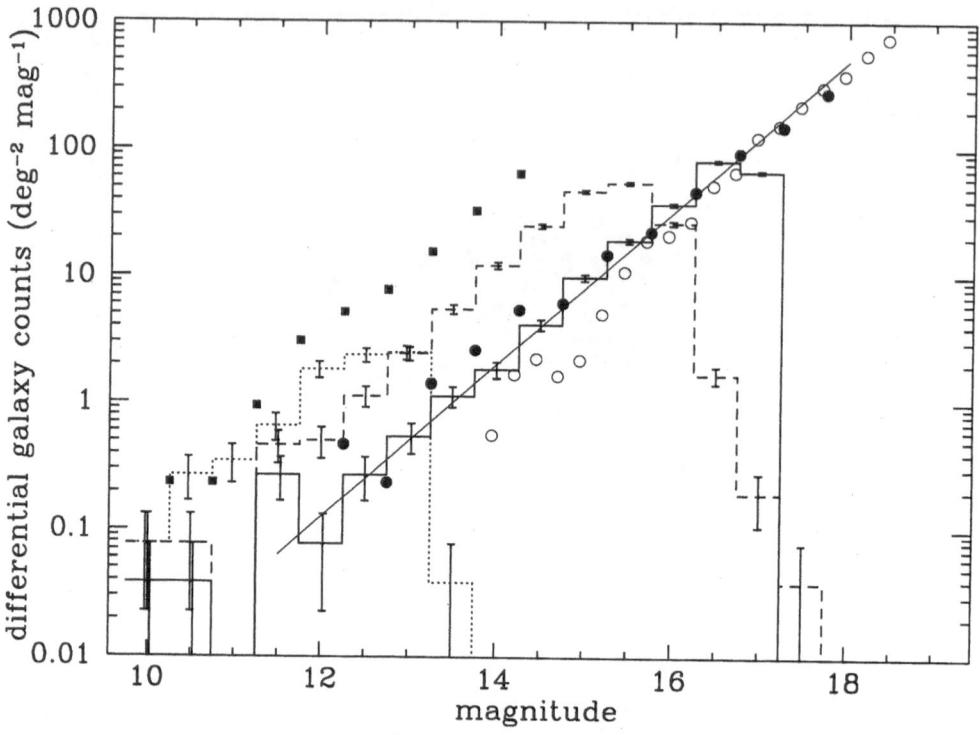

Figure 7. Differential galaxy counts over 50 deg^2 of high galactic latitude ($20° < |b|60°$), $I < 17$ DENIS data. Objects closer than 20 pixels to the frame edges are excluded. *Solid, dashed,* and *dotted histograms* refer to the DENIS I, J, and K counts, respectively. The *squares* represent K-band counts from Gardner et al. (1996), while the *open* and *filled circles* represent the I-band counts from Lidman and Peterson (1996), and Gardner et al. (1996), respectively. The *thin oblique line* represents an eye-fit Euclidean count function (0.6 slope).

TABLE 2. Estimated DENIS limits from 50 deg^2 of reduced data

	I_c	J	K
completeness ($\simeq$ 80%)	17.25	15.25	11
star/galaxy separation (90% reliability, from I)	16.5	14.8	13.5
photometry (0.20 mag accuracy)	>18?	14.8	12.2?
photometry (0.10 mag accuracy)	17.4	13.7	<11?

Using the counts from Figure 7 to extrapolate to the entire survey area (roughly a hemisphere), we infer that our homogeneous catalogs will have sizes of 6000 at $K < 11$ (0.2 mag photometry), 100 000 to 500 000 at $J <$

13.7 and 14.8 (with 0.2 and 0.1 mag photometry, respectively), and 900 000 galaxies at $I < 16.5$ (0.1 mag photometry). The recent installation of an air conditioning system on the K band optics has decreased the instrumental background by 0.7 magnitude, which should bring the extraction limit to $K \simeq 11.7$, and thus increase the size of the homogeneous K sample to roughly 15 000 galaxies.

Moreover, there is still room for progress on star/galaxy separation. C. Alard has devised a new algorithm to accurately model the variations of the PSF across the frame, which need no longer be an elliptical gaussian (fitting the asymmetric coma of the images), and tests on visually classified data are about to be performed.

Acknowledgements

We thank Emmanuel Bertin for supplying recent updates of his SExtractor software package, Steve Maddox for useful comments on the APM data, Nicolas Epchtein for a careful reading of the manuscript, and Pascal Fouqué and the DENIS operations team.

References

Baugh C.M. & Efstathiou G. (1993) *MNRAS*, **Vol. no. 265**, pp. 145–332
Baugh C.M. & Efstathiou G. (1994) *MNRAS*, **Vol. no. 267**, pp. 323–332
Bertin E. (1996) *PhD thesis*, University of Paris 6
Bertin E. & Arnouts S. (1996) *A&AS*, **Vol. no. 117**, pp. 398–404
Bertin E. & Dennefeld M. (1997) *A&A*, **Vol. no. 317**, pp. 43–53
Borsenberger J. (1997) in *The Impact of Large-Scale Near-IR Surveys*, eds F. Garzón *et al.*, Kluwer, pp. 181–186
Burstein D. & Heiles C. (1982) *AJ*, **Vol. no. 87**, pp. 1165–1189
Dalton G.B., Maddox S.J., Sutherland W.J. & Efstathiou G. (1997) *MNRAS*, **Vol. no. 289**, pp. 263–284
Epchtein N. *et al.* (48 authors) (1997) *ESO Messenger*, **Vol. no. 87**, pp. 27–34
Escalera E. & MacGillivray H.T. (1995) *A&A*, **Vol. no. 298**, pp. 1–21
Escalera E. & MacGillivray H.T. (1996) *A&AS*, **Vol. no. 117**, pp. 519–555
Gardner J.P., Sharples R.M., Carrasco B.E. & Frenk C.S. (1996) *MNRAS*, **Vol. no. 282**, pp. L1–L6
Heydon-Dumbleton N.H., Collins C.A. & MacGillivray H.T. (1989) *MNRAS*, **Vol. no. 238**, pp. 379–406
Kolatt T., Dekel A. & Lahav, O. (1995) *MNRAS*, **Vol. no. 275**, pp. 797–811
Kraan-Korteweg R.C., Woudt P.A., Cayatte V., Fairall A.P., Balkowski C. & Henning P.A. (1995) *Nature*, **Vol. no. 379**, pp. 519–521
Kron R.G. (1980) *ApJS*, **Vol. no. 43**, pp. 305–325
Lidman C.E. & Peterson B.A. (1996) *MNRAS*, **Vol. no. 279**, pp. 1357–1379
Lumsden S.L., Nichol R.C., Collins C.A. & Guzzo L. (1992) *MNRAS*, **Vol. no. 258**, pp. 1–22
Maddox S.J., Sutherland W.J., Efstathiou G., Loveday J. & Peterson B.A. (1990) *MNRAS*, **Vol. no. 247**, pp. 1P–5P
Mamon G.A. (1994) *Astrophys. & Sp. Sci.*, **Vol. no. 217**, pp. 237–242
Mamon G.A. (1996) in *Dark Matter in Cosmology, Quantum Measurements, Experimen-*

tal Gravitation, eds. R. Ansari, Y. Giraud-Héraud & J. Trân Thanh Vân, Frontières, pp. 225-232 (astro-ph/9608076)

Mamon G.A., Banchet V., Tricottet M. & Katz D. (1997a) in *The Impact of Large-Scale Near-IR Surveys* eds. F. Garzón *et al.*, Kluwer, pp. 239-248 (astro-ph/9608077)

Mamon G.A., Tricottet, M., Bonin, W. & Banchet, V. (1997b) in *Extragalactic Astronomy in the Infrared*, eds. G.A. Mamon, T.X. Thuan & J. Trân Thanh Vân, Frontières, pp. 369-380 (astro-ph/9711281)

Monet D. (1997) *Bull. A.A.S.*, **Vol. no. 188**, 54.04

Prandoni I., Iovino A. & MacGillivray H.T. (1994) *AJ*, **Vol. no. 107**, pp. 1235-1244

Saunders W. et al. (15 authors) (1997), in *Extragalactic Astronomy in the Infrared*, eds. G.A. Mamon, T.X. Thuan & J. Trân Thanh Vân, Frontières, pp. 415-424

Schröder A., Kraan-Korteweg R.C., Mamon G.A. & Ruphy S., (1997) in *Extragalactic Astronomy in the Infrared*, eds. G.A. Mamon, T.X. Thuan & J. Trân Thanh Vân, Frontières, pp. 381-386 (astro-ph/9706093)

Vauglin I., Paturel G., Marthinet M.C., Petit C. & Borsenberger J. in *Extragalactic Astronomy in the Infrared*, eds. G.A. Mamon, T.X. Thuan & J. Trân Thanh Vân, Frontières, pp. 387-392

EXTRAGALACTIC ASTRONOMY WITH 2MASS

S. E. SCHNEIDER AND J. L. ROSENBERG

UMass Astronomy Program, Amherst, MA 01003 USA

T. H. JARRETT AND T. J. CHESTER

IPAC, Caltech 100–22, Pasadena, CA 91125 USA

AND

J. P. HUCHRA

Harvard–Smithsonian CfA , Cambridge, MA 02138 USA

1. Introduction

Observations for the Two–Micron All Sky Survey (2MASS) have recently begun at the northern hemisphere site. The project is now in the final phases of tuning the data analysis software in preparation for running the "pipeline" at full bore. The path to completing the extragalactic portion of the pipeline, called *GALWORKS*, was complicated. It required an enormous number of ideas, compromises, and decisions to generate as accurate and as useful a set of data as we possibly can from the raw survey data.

At this meeting and the previous ones, we have described some of the particular software developments for extragalactic astronomy with 2MASS (Chester & Jarrett 1995; Schneider et al. 1997; Jarrett et al. 1997; Jarrett 1998; Rosenberg 1998). In this paper, we will review these developments more generally. By reviewing *GALWORKS* and its development, we can shed some light on how it came to take its particular final form, and perhaps aid future users of 2MASS data.

To put the 2MASS *GALWORKS* project in context, we begin with a brief review of earlier cataloging projects. We then turn to the goals and implementation of the survey, and consider a number of the issues involved in designing the software. Finally, we give some initial results for galaxy counts from the 3–channel survey camera.

193

N. Epchtein (ed.),
The Impact of Near-Infrared Sky Surveys on Galactic and Extragalactic Astronomy, 193-208.
© 1998 *Kluwer Academic Publishers.*

2. A Brief History of Galaxy Cataloging

Actually, although most of our discussion focuses on galaxies, the 2MASS "galaxy" catalog will include all sources that are detectably extended. Early extended-source catalogs included a large fraction of Galactic sources: 32% of the 103 sources in Messier's (1781) list are galaxies, approximately 85% of the New General Catalog/Index Catalog (Dreyer 1888, 1895, 1908) objects are galaxies. At the faint levels 2MASS will reach, we expect the extended source counts will be overwhelmingly dominated by extragalactic sources. It will certainly include many nearby extended sources like planetary nebulae and HII regions, but we do not have enough experience at low Galactic latitudes yet to predict how many of these sources we will identify.

We face almost identical problems as the earliest extended source catalogs. Like Messier, we have to worry about accidental duplication (M102 = M101), double stars (M40), tight groupings of a few stars (M73), and spurious sources (M91). (Note that today M102 is sometimes assigned to NGC 5866 and M91 to NGC 4548.) In modern parlance, Messier's list was about 96% reliable. Messier also "missed" seven sources which subsequent observers felt should be added to bring up the overall list to 110. Thus he was 93% complete. Our targets for the 2MASS extended source completeness and reliability are actually quite similar as we discuss later.

The completion of the Palomar Observatory Sky Survey (POSS) in the 1950's marked a new degree of uniformity in sky coverage. A number of new galaxy catalogs soon followed based on direct inspection of the POSS Schmidt plates: The Catalogue of Galaxies and of Clusters of Galaxies carried out by Zwicky et al. (1961–1968) for 31,000 galaxies brighter than $B \approx 15.7$ mag in the northern hemisphere; The Morphological Catalog of Galaxies (Vorontsov–Velyaminov et al. 1962–1968), a somewhat less uniform survey of 29,000 galaxies north of $\delta = -33°$; and the Uppsala General Catalogue (Nilson 1973) of 13,000 northern galaxies, which is diameter limited to about $1'$. In addition, Lauberts (1982) generated a catalog of 18,000 southern galaxies ($\delta < -18°$) from the ESO(B) Schmidt plates. These surveys remain the primary all-sky optical catalogs of galaxies even though the subjective aspect of human inspection of the images leaves much to be desired.

In recent years galaxy cataloging has entered a transitional period from photographic to digital data, and from subjective "eyeball" to more objective computer software schemes. In certain ways these transitions are linked, since the nonlinear response of photographic plates has many subtleties that are difficult to define in terms of a set of straightforward computer algorithms. Star/galaxy separation has become automated by using automatic plate scanning machines like COSMOS and APM (see, for example, Maddox et al. 1990) and developing classification methods like FOCAS (Jarvis & Tyson 1981). These approaches define a variety of parameters for each source, often keyed to the idea that at a given magnitude an extended source will cover a larger area of the image (to some fiducial surface brightness limit) and will have a smaller peak flux density. A variety of other measures like 2nd, 3rd, 4th moments, degree of elongation, and measures of asymmetry or multiple peaks (to exclude multiple stars) were added to further aid in discrimination. These were compared to the locus of parameters for stars as a function of magnitude, and a measure of the deviation from being star-like was used to set the catalog boundaries.

In recent years, neural network schemes (Odewahn et al. 1992) and decision tree methods (Weir et al. 1995) have been developed that can more effectively adapt to unusual circumstances like seeing, plate sensitivity, and image artifacts. While the approaches are somewhat different, both use training sets to "teach" the algorithm how to recognize galaxies. These approaches identify the combinations of parameters that best distinguish between the different types of real and artificial sources, and can even be used to classify different morphological types of galaxies. A drawback, however, is that the flexibility of the selection criteria can make it difficult to understand the completeness of the sample.

The 2MASS extended source catalog is based on a rigid star/galaxy separator primarily. It is probably fair to say that the need for neural networks and decision trees grows in direct proportion to the messiness of the raw data. Thus photographic surveys are especially well served by them. Because the 2MASS raw data are so uniform, we believe we can meet survey goals without them. Still, close to the flux limit of the survey, it may be possible to extract significantly more useful data, and we are currently investigating a decision tree approach to see what results it yields.

We do not currently plan to attempt any morphological classifications for the galaxies, in part because the cues for these distinctions are much weaker in the infrared. Actually, such parameters as color differences may be powerful discriminators between galaxy types (and between other non-galaxian sources), but we do not plan to build those kinds of distinctions into the catalog other than to provide the basic data.

3. 2MASS Extragalactic Objectives

Even before the final design of the 2MASS telescopes and cameras, we set a number of goals for the extragalactic parts of the survey. These Level–1 specifications are listed in Table 1, and were based on early assessments of detector sensitivity, telescope size, and survey duration. The flux limits were estimated on the basis that we ought to be able to detect an extended source distributed over roughly twice the diameter of a seeing-convolved point source. Thus the integrated background noise would be approximately two times worse, which would allow us to detect a source about 0.8 mag brighter than the point source limit.

TABLE 1. Level–1 Specifications for Extended Sources

Sensitivity:	J<15.0 H<14.2 K_s <13.5
Photometric Uniformity:	< 10%
Photometric Precision:	< 10% (for H<13.8)
Reliability:	> 99% ($|b| > 20°$)
	> 80% ($|b| > 10°$)
Completeness:	> 90% ($|b| > 30°$)
	(for galaxies with scale lengths > 0.5'')

These photometric limits allow us to make some quick estimates of the likely number of sources we should detect based on the optical catalogs. Assuming an average color of $B - K_s = 4$, the K_s–band sensitivity limit of 13.5 is comparable to an optical limit of ~17.5. The CGCG includes ~31,000 galaxies in northern hemisphere to about 1.8 mag brighter than this, so there would be a factor of $\sim 12\times$ more galaxies. Correcting for the whole sky and zone of avoidance suggests that roughly a million galaxies will be detected by 2MASS. We will revisit this estimate at the end of the paper.

It is also important that the galaxy photometry be adequate for the kind of science one might envision doing with a near-infrared survey. One of the more important directions for follow-up studies will be to use the infrared Tully–Fisher relation (IRTF) to determine galaxy distances (see Rosenberg 1998). Since the IRTF is at best accurate to about 15% (see, for example, Willick et al. 1996), the survey photometry and calibration should be more accurate than this.

The specification on calibration uniformity is 10% over the whole sky. We expect to maintain at least this degree of uniformity by frequent re-observation of the same set of calibration fields, by comparison of stellar photometry in overlap regions of the survey, and by comparisons to observations made with other instruments in an assortment of fields chosen to represent a wide range of Galactic longitudes and latitudes. One of the biggest concerns is that the photometry be well matched between the northern and southern telescopes. We hope to assure this by observing the same calibration fields along the equator, by overlapping the coverage at the boundary between north and south, and by cross–observing selected fields in the other telescope's hemisphere.

The photometric precision required for repeated observations of the same source is also set at 10%, but for sources about half a magnitude brighter than the sensitivity requirement. The brighter sources will also generally be the larger sources for which other important parameters for the IRTF like the inclination will be well measured. The specification is listed in terms of the H–band sensitivity since, at the time, the H–band IRTF was favored. It appears that K_s–band may be as good or better, and we expect to have a similar precision for galaxies half a magnitude brighter than the limiting sensitivity at K_s as well.

Finally, we set completeness and reliability values for the galaxy catalog. We cannot detect some galaxies that meet the catalog's nominal magnitude cutoff because they are too small, too nucleated, or too low in surface brightness. We try to summarize such problems by a scale-length requirement, which specifies a fraction of the flux that should extend into the wings of the PSF to make a source detectably extended. Based on repeated observations of the same field and by comparisons to other infrared and optical images of the same fields, we appear to be meeting the completeness requirement of 90% quite readily. The reliability limit of 99% is a much greater challenge, especially at Galactic latitudes approaching 20°, where the number of double and triple stars, which can look like extended sources, grows large. Much of the final tuning of the *GALWORKS* pipeline is directed at this problem.

4. Hardware Requirements and Limitations

The ability to carry out a near infrared sky survey has been provided by the development of NICMOS–3 detector arrays for the Space Telescope. These 256×256 pixel arrays provide excellent sensitivity to beyond 2μ, and allow us to reach sufficiently faint levels for galaxy cataloging in short integrations even with a relatively small telescope. The extragalactic goals of the 2MASS project place some serious requirements that ultimately affect the duration (and cost) of the survey: (1) we need to sample the sky on scales small compared to the point spread function, so that stars and galaxies can be separated; and (2) we need long enough integrations to provide surface brightness sensitivity to the outer parts of galaxies.

The compromise we have reached, which allows the survey to be completed within about 3 years, is to design the optics to have a plate scale of $2''$ per pixel and integrate about 8 seconds on each point of the sky. The $2''$ pixel size is not ideal for star/galaxy separation, but by collecting the data in six 1.3 s integrations, offset by a fraction of a pixel ("dithering"), an adequate sampling is achieved. The total integration is sufficiently long to detect K_s surface brightnesses of about 20 mag arcsec^{-2}. This value sounds surprisingly faint compared to the K_s–band point source sensitivity of 14.3, but it should be kept in mind that that is a 10–σ precision limit for point sources that may be spread over up to four $2'' \times 2''$ pixels. The rms noise in an individual pixel is closer to 19 mag arcsec^{-2}, and for isophotal fits to extended sources, many pixels can be averaged. The resulting images are close to the sensitivity of the POSS when the color differences in galaxies are taken into account. Fig. 1 shows some sample galaxies detected with 2MASS compared to their blue images as found on the digitized POSS.

Unlike DENIS, the 2MASS frames are dithered "on the fly." As the telescope swings along fixed-R.A. scans, a slight tilt of the detector array shifts the image center a fraction of a pixel to the side from frame to frame. (See Fig. 2.) Each frame is frozen for 1.3 s by a tip–tilt secondary mirror, and the exact timing determines the fractional pixel offset along the direction of the scan. By an appropriate combination of R.A. and Dec. offsets, for example 1/6th of a pixel in R.A. and 42 2/3rd pixels in Dec., the pixel centers can be quite evenly spaced on the sky, providing good PSF sampling. The actual offsets chosen are slightly complicated by the need to provide good sampling for all three of the J, H, and K_s arrays, which do not have precisely the same plate scale.

Since galaxies are extended, and since measuring their fluxes and sizes across frames is difficult and inaccurate, the extragalactic portion of 2MASS also puts a requirement on frame overlap. The in-scan frames overlap by 5/6ths, and a continuous image 8.5' wide by 6° in declination is generated. The scans are overlapped by 10% (0.85') in R.A. so that sources of this size and smaller will be entirely on a single frame. We also measure galaxies which are at least 75% on the frame, so that we will include all galaxies up to about 1.7'. These galaxies should have fairly accurate parameters since most of the light comes from the center, and the blocked outer portions will be corrected by isophotal substitution based on the fits to the interior regions.

This leaves a fairly small number of large galaxies out of the catalog. The UGC has 3465 galaxies 1.7' or larger in the northern hemisphere. Assuming these galaxies fall randomly on the array, only about 400 would fall so close to the edge that less than 75% was on either of two neighboring scans. Over the whole sky this number would double, and correcting for zone of avoidance effects might double it again. Thus the catalog should be incomplete for only the largest ~14,000 galaxies, and of these, only ~1600 will be caught so close to the scan edge that they cannot be measured.

The overlap between scans creates a different problem. Between 5 and 10% of the sources will be duplicated. One of the reasons we identify the

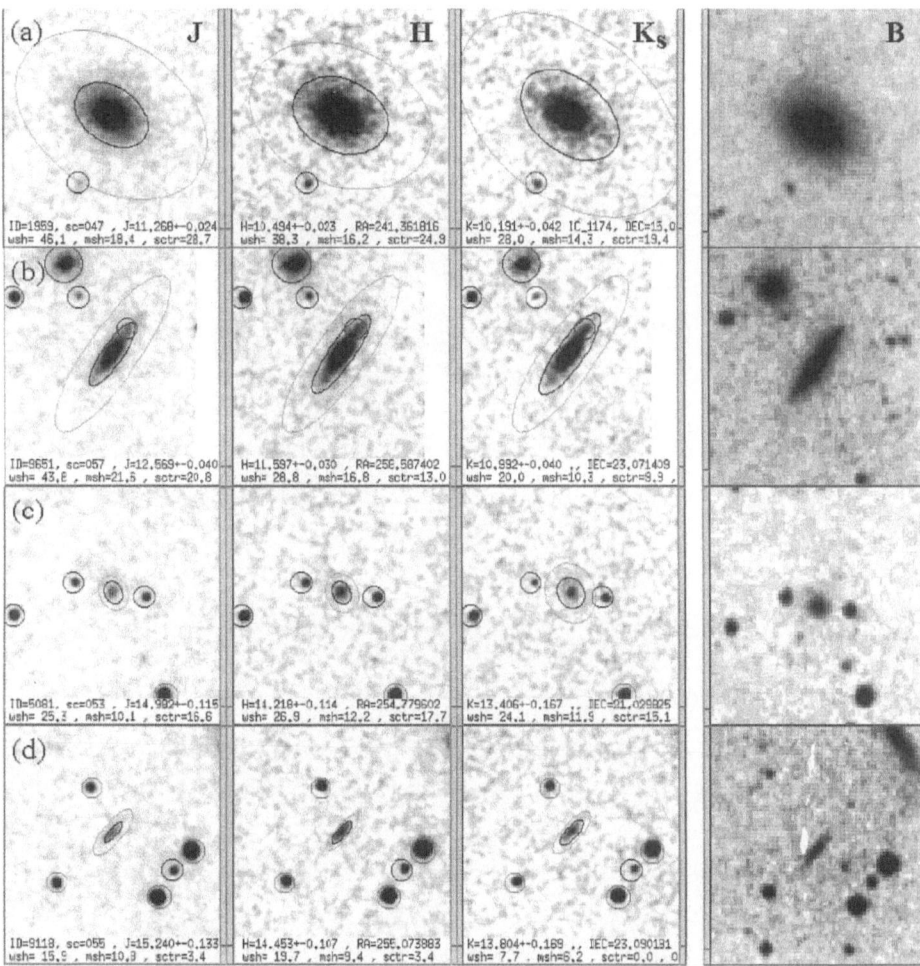

Figure 1. An assortment of galaxies detected by 2MASS. Each galaxy is centered in a 100″ × 100″ box. The dark ellipse surrounding each galaxy indicates where the surface brightness has dropped to 20 mag arcsec^{-2}. The fainter ellipse is a Kron elliptical aperture. Small circles surround other sources that were blanked from the image. The top two objects are examples of "bright" galaxies. Galaxy (a) is IC 1174, with $K_s = 10.2$, and a photographic magnitude of 14.5. Galaxy (b) is optically uncataloged at $K_s = 11.0$. The bottom two galaxies are close to the nominal sensitivity limit of the survey. Galaxy (c) has $K_s = 13.4$; galaxy (d) at $K_s = 13.8$ demonstrates that edge-on galaxies can be clearly detected to beyond the 2MASS sensitivity limit.

location of sources by their brightest point (as opposed to a centroid) is that it is a uniquely defined position. The centroid of a source that is cut off at the edge of a frame will be quite different from one frame to the next, making it difficult to cross–identify.

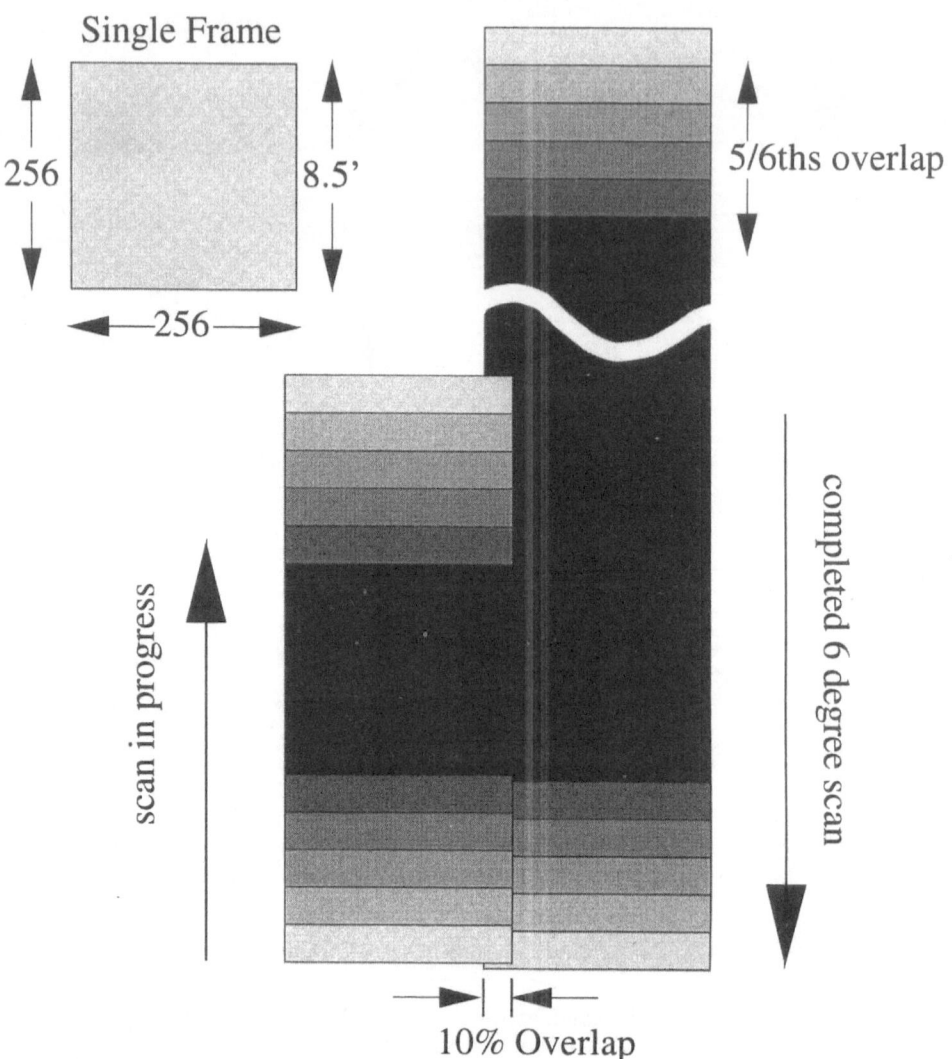

Figure 2. 2MASS is constructed from a series of 256×256 image frames strung together in 6°-long declination scans. The frames are overlapped by 5/6ths so that each spot on the sky is observed for 1.3 s six times. The scans are overlapped by 10% so that galaxies as large as 0.85′ are always contained on a single scan.

5. Extended Source Identification

The method for identifying galaxies from the 2MASS data was developed during experimentation with prototype camera images. We tested various parameters like those discussed in section 2 to find the ones that did the best job of separating stars and galaxies. The problem is somewhat more difficult for a near-infrared survey like 2MASS because the effective sensitivity is not quite as good as that of optical surveys, and in particular, the blue color of spirals' disks make them weaker relative to the nucleus.

We began by examining several fields, particularly in the direction of galaxy clusters to study which parameters best discriminated between stars and galaxies. These results were based on a comparison to optical determinations from the POSS. The parameters that proved the most powerful in broadly identifying extended sources were a basic maximum surface brightness vs. magnitude comparison ($MXDN$), and a more complicated "radial shape discriminant" parameter (SH) based on fitting a generalized exponential to the distribution. This latter parameter is determined by fitting to the azimuthally averaged radial distribution defined by:

$$f(r) = f_0 e^{-(r/\alpha)^{1/\beta}} \tag{1}$$

where α is a scale length and β is the rate of exponential decay. These parameters allow the distribution to take on a variety of forms ranging from Gaussian ($\beta = 1/2$) to an exponential disk ($\beta = 1$) to a de Vaucouleurs $r^{1/4}$ law ($\beta = 4$). Our original hope was that this might also prove to be useful for determining morphological types of galaxies, but perhaps because spirals are bulge dominated in the infrared, this does not appear to be the case. In fact, α and β tend to be strongly correlated, so we have defined a single parameter (SH) as the product $\alpha \times \beta$, which ranges from small values for stars to larger ones for galaxies.

The camera optics provide an essentially uniform PSF across the arrays, but one thing that became clear in the testing phase of 2MASS was that the star/galaxy separation depends critically on the seeing. Thus we cannot choose fixed cutoff values for the star/galaxy separation parameters because that would either reject too many galaxies when the seeing is good or accept too many stars when the seeing is poor. In fact, the star/galaxy separation depends on knowing the seeing to better than 0.1″. This at first appeared to be a serious problem, but we discovered we could reverse the situation by using the shape parameter to track the seeing. We use the large number of stars—even at high Galactic latitude and even while passing through a galaxy cluster—to provide us with a large number of point sources which all have consistently low SH scores that depend only on the magnitude and seeing. Then, using a running average over a timescale of minutes, we determine which sources are extended based on their score relative to the lower envelope of SH scores.

On the initial pass, we accept sources that pass a very weak SH or $MXDN$ test, then we tighten up the requirements with a set of tests to re-

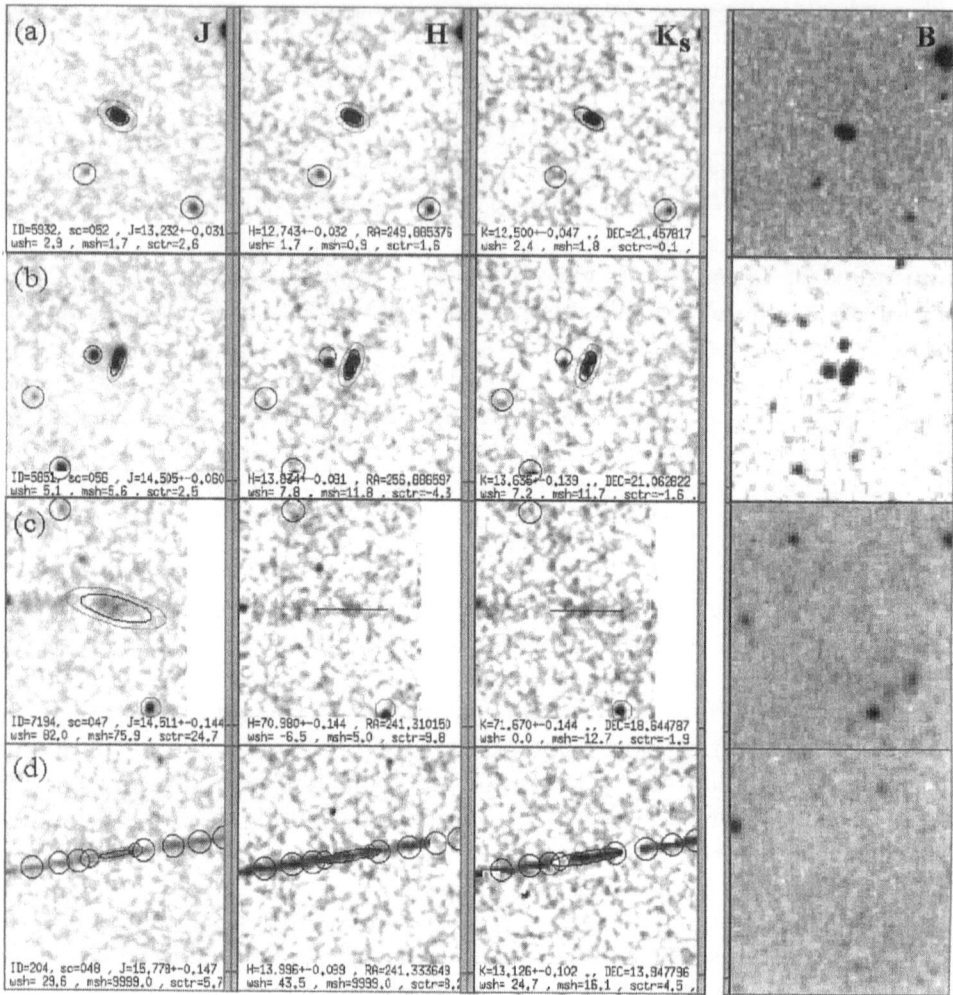

Figure 3. Examples of false extended sources. Images are displayed as in Fig. 1. Source (a) is a double star, and (b) is a triple star. Both have values in the modified shape tests that allow them to be distinguished from galaxies. Object (c) is an unmasked diffraction spike of a star. It was picked up in the J–band image, probably because of a faint star that is superimposed, but undetected at H and K_s. Object (d) is a portion of a meteor trail. Despite its obvious appearance to the eye, these trails have proved "expensive" to detect in *GALWORKS*.

ject various false sources. The primary contaminants are double and triple stars with increasing frequency at lower Galactic latitudes. We also have to struggle with meteor trails and diffraction spikes from bright stars, especially when the star is off the frame. (See Fig. 3.)

An assortment of parameters are being used to identify and reject false sources. Several of our tests examine the shape score over limited portions of a source in order to eliminate the contribution from isolated secondary sources. For example, we can examine the values of the shape parameter in wedge-shaped "pie slices" (where the vertex of the wedge is centered on the point of peak flux density). A galaxy should show a consistently large SH score in each wedge, but a binary star will have small scores except in the wedge that contains the secondary star. Similarly, we can find the SH score for the *median* flux level in annuli at each radius. Again, this should not change much for a galaxy, but it will tend to eliminate contributions from secondary sources that contribute a significant amount of light over less than half of the azimuthal distribution about a source.

Our current strategy is to accept potential sources to about a magnitude fainter than the Level–1 specifications in any of the bands. These sources become part of a working survey database, which we then pare back to generate the "official" database based on a much larger set of acceptance and rejection criteria that we carry along with each source (including several that we are not currently implementing). This allows us to experiment with modifications to the galaxy selection procedure without having to rerun the extended source processor on the raw data.

6. Magnitude and Surface Brightness Issues

Up to this point we have cheated a little by talking about galaxy magnitudes as though they were well defined. One of the major problems in galaxy cataloging is the difficulty in measuring anything remotely resembling a "total flux." Even with effective surface brightness limits substantially fainter than in 2MASS, total magnitudes can only be approximately extrapolated from magnitudes measured within fixed bright isophotes, and this introduces large errors. Isophotal magnitudes, on the other hand, may be more consistent, but they are somewhat arbitrary. This is part of the reason that there is such a proliferation of techniques for measuring galaxy magnitudes.

Since galaxies fade to unmeasurable surface brightnesses while still missing a significant fraction of their total of light, we are forced to select between methods that are accurate but imprecise, and others that are precise but inaccurate. For example, circular apertures are less likely to introduce biases due to the exigencies of ellipse fitting. They also should introduce a similar level of faint star contamination in edge-on versus face-on galaxies. However, elliptical apertures, having less contamination and a smaller background noise contribution will tend to be more precise. For statistical studies, it is better to be accurate, but in studies of an individual galaxy where color comparisons are being made, precision is better.

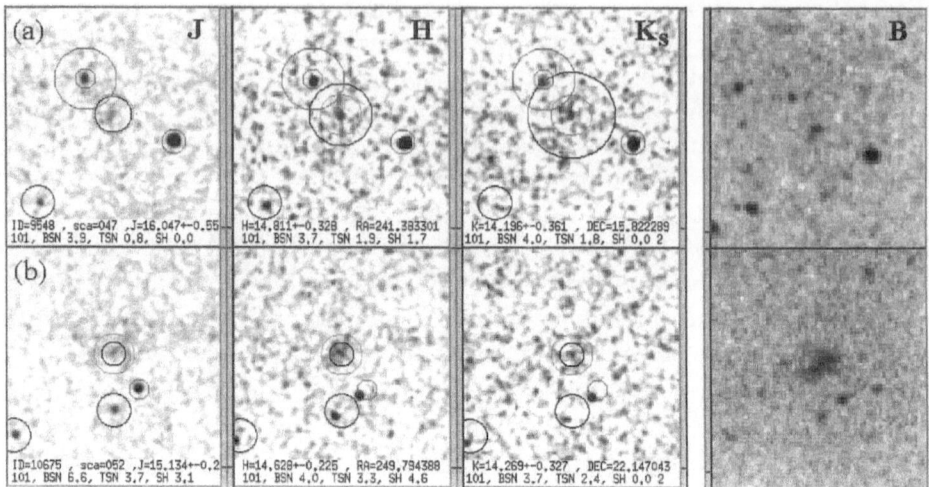

Figure 4.   Low central surface brightness galaxies detected by Algorithm 2. Plotted as in Fig. 1. At K_s = 14.2 and 14.3, these objects are outside of the sensitivity limit requirements, but they demonstrate the potential of the survey for deeper "mining."

These conflicting needs are reflected in the 2MASS database. We are recording a large assortment of magnitudes in the working survey database, although only a few of them are likely to appear in the released catalog. Currently we determine magnitudes in a series of fixed circular apertures, isophotal magnitudes for circular and elliptical isophotes, Kron magnitudes as is done in DENIS, and Petrosian magnitudes. The Petrosian magnitudes are determined by finding the isophote at a fixed fraction of the mean surface brightness interior to the isophote. Such a magnitude is unbiased in principle, but it has proved very imprecise, possibly because of the limited dynamic range of our survey. In principle, we could determine each kind of magnitude in each band, and then find the corresponding magnitude in the other bands, but we have decided to treat the K_s–band as the fiducial as much as possible since this is foremost a 2μ survey.

Another concern for any imaging survey is the problem of overlooking sources with large integrated fluxes whose light is too broadly distributed to be picked out at the early stages of image processing. For example, the 2MASS extended source processor begins by looking around local maxima in the images. Since most galaxies are strongly centrally condensed, this is a reasonable operative approach, but we know of some galaxies that never exceed the local maximum requirement, yet have a large integrated flux.

To search for these objects, we have developed a second algorithm to examine the images after first subtracting point and extended sources detected earlier in the pipeline. This "Low Central Surface Brightness"

(LCSB) processor smooths the residual images and searches for significant peaks. This process picks up some surprisingly faint sources (see Fig. 4), but it does not appear to add a large enough number to indicate that our Level–1 specification on completeness might otherwise be compromised. We plan to continue running this processor, however, since it detects some very interesting sources. The LCSB processor is discussed in much more detail by Jarrett (1998) elsewhere in this volume.

7. Image Masking and Blanking

Even though the 2MASS data are very uniform, there are an assortment of situations that force us to excise data in various ways. In some cases we need to mask out data from any consideration in the extended source processor, while in others we need to manipulate the image, for example, blanking out stars that overlay a galaxy.

One of the consequences of our stringent reliability criterion is that we need to avoid regions that are likely to generate false extended sources. One such area is in the wings of the PSF of bright stars. Bright stars also generate bright diffraction spikes, persistence "ghosts" that may take up to 10 s to fade from the image, and "horizontal stripes," an electronic read-out problem in rows offset by 128 pixels from a bright star. All of these artifacts represent a sort of high-dynamic-range PSF, which is not well enough understood to allow us to simply subtract it from the data. This is frustrating, because we have already encountered many examples of fairly bright galaxies that were clearly visible, but lay on these features, making them unreliable. The amount of area masked can also grow very large as the star density increases near the Galactic plane.

To minimize these problems, we are trying to determine worst-case levels for these various features relative to the brightness of the source so that we can mask them only where necessary. Our goal is to establish the highest safe level for these features relative to the image background noise. The background noise includes an estimate of the effective noise caused by stellar confusion so that deep in the Galactic plane where the bright star artifacts are swamped by other stars, we do not mask out regions that might prevent us from detecting a bright galaxy.

Large galaxies and HII regions also present a problem for the extended source processor, which tends to break them up into a large number of smaller sources. This is another annoying problem, although it includes a relatively small number of sources. Our strategy is to use outside catalogs to flag such regions. For the largest of these regions we simply turn off the processing because it becomes extremely inefficient, and "expensive" computationally speaking. For other cataloged sources we merely flag the source or sources that are detected within its boundaries.

For the data that survives the blanking process, there are still sources present that interfere with making isophotal fits and magnitude measurements. We identify sources that are within about 1 mag of the galaxy's brightness. These are then blanked out for isophotal fitting procedures, and an isophotal substitution is performed before determining magnitudes.

This procedure was debated before the survey began because there is a risk that nearby, large galaxies will be handled differently than poorly-resolved distant galaxies. For example, an obviously-extended HII region in a nearby galaxy might be treated as a contaminating star if the galaxy were farther away. Similarly, it might seem more accurate to determine the local background level in an annulus surrounding a galaxy *without* removing the stars. This is because it is much more difficult to identify stars against the galaxy itself, so leaving the stars in the annulus biases it in the same way. In the prototype camera tests, though, we found that if we did not work on removing stars, the magnitudes simply became unacceptably imprecise, and they did not show any significant improvement in the biasing.

8. The Real Beginning

As of this writing (October 1997), 2MASS has already surveyed 7% of the northern sky. We expect a fairly large fraction of this data to survive the quality-assurance tests and to become part of the final survey data. The software is essentially complete, except for some parameter tuning, and we are examining fields so that we can reach some final decisions on the best choices for these parameters.

Based on our work to this point, we have offered a quick look at some of the individual source images, and how they are handled. We are also now in a better position to say something about the probable size of the 2MASS catalog based on these early analyses. In Fig. 5, we plot the extended source counts from two essentially random regions at high Galactic latitudes. The regions cover about 6 and 10 sq deg after we have eliminated some apparent clusters in each.

The figure shows the differential source counts in half magnitude bins. The dashed line shows the simple volume increase, and indicates that the counts are complete to $K_s = 13.5$, and almost complete to $K_s = 14$. The counts in these regions indicate that there are about 26 sources per sq deg, or about 1.1 million sources over the whole sky to $K_s = 13.5$. This is close to our initial extrapolations from the optical catalogs. Our counts are in good agreement with the values found by Gardner et al. (1996) as is shown if Fig. 5.

These counts are slightly higher than values found with our prototype

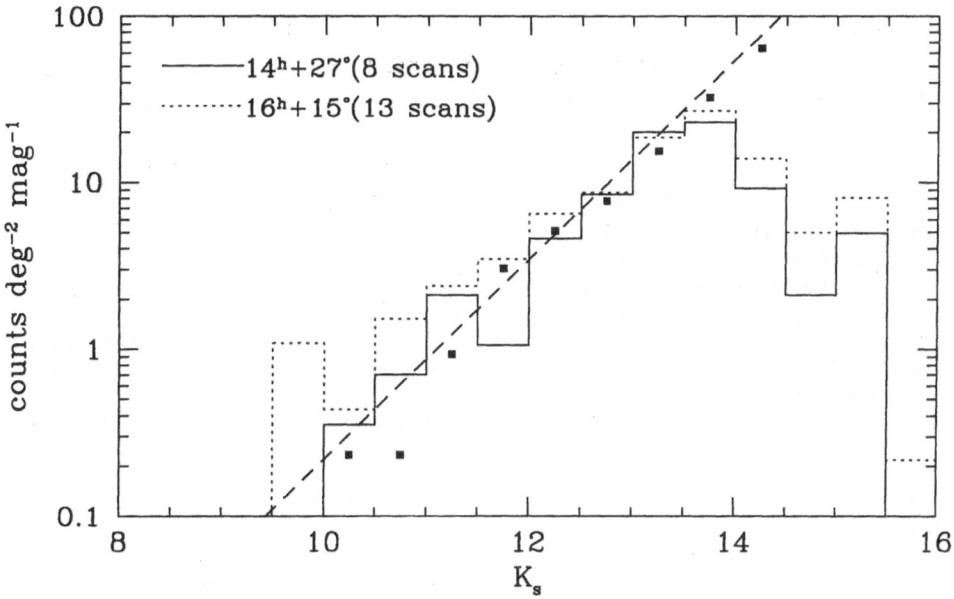

Figure 5. Galaxy counts drawn from two fields at high Galactic latitude. Counts are per half-magnitude bins in K_s, normalized to one mag interval. The dashed line shows the simple increase with volume of a complete sample, and the square symbols show the values found by Gardner et al. (1996) in an 8.5 sq deg region.

camera. The 3–channel camera is more successful at detecting galaxies close to and beyond our sensitivity limit, at least in part because it has three opportunities to catch them. We had been cautious in interpreting our earlier counts because we thought they might be *over*estimates since most of the prototype fields were targeted on rich galaxy clusters. Still, it was apparent in the redshift results from the Abell 262 region, for example, that most of the galaxies detected were not members of the cluster (Schneider et al. 1997). It is clear now that the total counts will be even higher.

Actually, if we do not exclude clusters, the total counts we will find will be significantly higher. We chose the fields in Fig. 5 at random, and they have no major clusters in them, but there were high density regions in each that would have raised the average counts by more than 50% in each case. (We note that Gardner et al. also excluded a cluster in their region.) If the regions we examined are representative, it seems likely that 2MASS may detect closer to 1.5 million extended sources with K_s <13.5. With sources also meeting the J and H sensitivity criteria, the 2MASS catalog will balloon to several million sources, and it seems clear that their will be several million more galaxies beyond our official completeness limit.

References

Chester, T. J., & Jarrett, T. 1995, Proc. Euroconference on Near Infrared Sky Surveys, San Miniato, Italy, eds. P. Persi et al., Mem. S. A. It., 66, 567

Dreyer, J. L. E. 1888, Mem R Astr Soc, 49, 1

Dreyer, J. L. E. 1895, Mem R Astr Soc, 51, 185

Dreyer, J. L. E. 1908, Mem R Astr Soc, 59, 105

Gardner, J. P., Sharples, R. M., Carrasco, B. E., & Frenk, C. S. 1996, MNRAS, 282, L1

Jarret, T. 1998, this proceedings.

Jarret, T., Chester, T., Schneider, S. & Huchra, J. 1997, in "The Impact of Large Scale Near-IR Sky Surveys," p. 213, F. Garzon et al. (eds.), Kluwer (Netherlands).

Jarvis, J. F., & Tyson, J. A. 1981, AJ, 86, 476

Maddox, S. J., Sutherland, W. J., Efstathiou, G., & Loveday, J. 1990, MNRAS, 243, 692

Messier, C. 1780, in "Connoissance des Temps pour l'Année bissextile 1784," p. 227, l'Imprimerie Royale (Paris).

Nilson, P. 1973, Uppsala General Catalogue of Galaxies, Uppsala Astr Obs Ann, 6

Odewahn, S. C., Stockwell, E. B., Pennington, R. L., Humphreys, R. M., & Zumach, W. A. 1992, AJ, 103, 318

Rosenberg, J. L. 1998, this proceedings.

Schneider, S. E., Huchra, J. P., Jarrett, T. H., & Chester, T. J. 1997, in "The Impact of Large Scale Near-IR Sky Surveys," p. 187, F. Garzon et al. (eds.), Kluwer (Netherlands).

Vorontsov–Velyaminov, B. A., et al. 1962, 1963, 1964, 1968, Morfologiceskij Katalog Galaktik, I–IV, Moscow State University (Moscow).

Weir, N., Fayyad, U. M., & Djorgovski, S. 1995, AJ, 109, 2401

Willick, J. A., Courteau, S., Faber, S. M., Burstein, D., Dekel, A., & Kolatt, T. 1996, ApJ, 457, 460

Zwicky, F., et al. 1961, 1963, 1965, 1966, 1968, Catalogue of Galaxies and of Clusters of Galaxies, I–VI, California Institute of Technology (Pasadena).

LARGE-SCALE STRUCTURES BEHIND THE MILKY WAY FROM NEAR-IR SURVEYS

R.C. KRAAN-KORTEWEG
DAEC, Observatoire de Paris, Meudon, France, and
Dept. de Astronomia, Universidad de Guanajuato, Mexico

A. SCHRÖDER
Institute of Astronomy, NCU, Chung-Li, Taiwan

G.A. MAMON
IAP, Paris, France, and
DAEC, Observatoire de Paris, Meudon, France

AND

S. RUPHY
DESPA, Observatoire de Paris, Meudon, France

Abstract. About 25% of the optical extragalactic sky is obscured by the dust and stars of our Milky Way. Dynamically important structures might still lie hidden in this zone. Various approaches are presently being employed to uncover the galaxy distribution in the Zone of Avoidance (ZOA) but all suffer from (different) limitations and selection effects.

We investigated the potential of using the DENIS NIR survey for studies of galaxies behind the obscuration layer of our Milky Way and for mapping the Galactic extinction. As a pilot study, we recovered DENIS I_c, J and K_s band images of heavily obscured but optically still visible galaxies. We determined the I_c, J and K_s band luminosity functions of galaxies on three DENIS strips that cross the center of the nearby, low-latitude, rich cluster Abell 3627. The extinction-corrected $I - J$ and $J - K$ colours of these cluster galaxies compare well with that of an unobscured cluster. We searched for and identified galaxies at latitudes where the Milky Way remains fully opaque ($|b| < 5°$ and $A_B \gtrsim 4 - 5^m$) — in a systematic search as well as around positions of galaxies detected with the blind H I-survey of the ZOA currently conducted with the Multibeam Receiver of the Parkes Radiotelescope.

N. Epchtein (ed.),
The Impact of Near-Infrared Sky Surveys on Galactic and Extragalactic Astronomy, 209-220.
© 1998 *Kluwer Academic Publishers.*

1. Introduction

Some of the results of this study have already been reported in Schröder *et al.* 1997 (Paper I). For a comprehensive description, the goals and earlier results of this project are repeated here, but the reader is referred to paper I for details on earlier presented results.

About 25% of the optically visible extragalactic sky is obscured by the dust and stars of our Milky Way. Dynamically important structures — individual nearby galaxies (*cf.* Kraan-Korteweg *et al.* 1994) as well as large clusters and superclusters (*cf.* Kraan-Korteweg *et al.* 1996) — might still lie hidden in this zone. Complete whole-sky mapping of the galaxy and mass distribution is required in explaining the origin of the peculiar velocity of the Local Group and the dipole in the Cosmic Microwave Background.

Various approaches are presently being employed to uncover the galaxy distribution in the ZOA: deep optical searches, far-infrared (FIR) surveys (*e.g.*, IRAS), and blind H I searches. All methods produce new results, but all suffer from (different) limitations and selection effects. Here, the near infrared (NIR) surveys such as 2MASS (Strutskie *et al.*, 1997) and DENIS in the southern sky, (Epchtein, 1997; Epchtein *et al.*, 1997) could provide important complementary data. NIR surveys will:

• be sensitive to early-type galaxies — tracers of massive groups and clusters — which are missed in IRAS and H I surveys,

• have less confusion with Galactic objects compared to FIR surveys,

• be less affected by absorption than optical surveys.

But can we detect galaxies and obtain accurate magnitudes in crowded regions and at high foreground extinction using NIR surveys? To assess the performance of the DENIS survey at low Galactic latitudes, we addressed the following questions:

(1) How many galaxies visible in the B_J band ($B_{\lim} \approx 19\overset{m}{.}0$) can we recover with DENIS in I_c (0.8μm), $J(1.25\mu$m) and $K_s(2.15\mu$m)? Although less affected by extinction (45%, 21% and 9% as compared to B_J), their respective limits for highly complete, reliable and photometrically accurate galaxy extraction with DENIS are lower ($16\overset{m}{.}5, 14\overset{m}{.}8$, and $\simeq 11$, Mamon *et al.* 1998).

(2) Can we determine the I_c, J, and K_s band luminosity functions?

(3) Can we map the Galactic extinction from NIR colours of galaxies behind the Milky Way?

(4) Can we identify galaxies at high extinction ($A_B > 4 - 5^m$) where optical surveys fail and FIR surveys are plagued by confusion?

(5) Can we recover heavily obscured spiral galaxies detected in a blind H I search and hence extend the peculiar velocity field into the ZOA via the NIR Tully–Fisher relation?

We pursued these questions by comparing available DENIS data with results from a deep optical survey in the southern ZOA (Kraan-Korteweg & Woudt 1994, Kraan-Korteweg *et al.* 1995, 1996, and references therein). In this region ($265° \lesssim \ell \lesssim 340°$, $|b| \lesssim 10°$), over 11 000 previously unknown galaxies above a diameter limit of $D = 0\overset{.}{'}2$ and with $B \lesssim 19\overset{m}{.}0 - 19\overset{m}{.}5$ have been identified (*cf.* Fig. 1 in Paper I). Many of the faint low-latitude galaxies are intrinsically bright galaxies. Within the survey region, we investigated DENIS data at what seems to be the core of the Great Attractor (GA), *i.e.*, in the low-latitude ($\ell = 325°$, $b = -7°$), rich cluster Abell 3627, where the Galactic extinction is well determined (Woudt *et al.*, 1997), and in its extension across the Galactic Plane where the Milky Way is fully opaque.

2. Expectations for DENIS galaxy extraction in the ZOA

What are the predictions for DENIS at low latitudes? In unobscured regions, the density of galaxies per square degree is 110 in the blue for $B_J \leq 19\overset{m}{.}0$ (Gardner *et al.*, 1996), and 30, 11, and 2 in the I_c, J and K_s bands for their respective completeness limits of $I_{\lim} = 16\overset{m}{.}0$, $J_{\lim} = 14\overset{m}{.}0$, $K_{\lim} = 12\overset{m}{.}2$ (see Mamon *et al.* 1997b). The number counts in the blue decrease with increasing obscuration as $N(< B) \simeq 110 \times \text{dex}(0.6\,[B - 19])\,\text{deg}^{-2}$. According to Cardelli *et al.* (1989), the extinction in the NIR passbands are $A_{I_c} = 0\overset{m}{.}45$, $A_J = 0\overset{m}{.}21$, and $A_{K_s} = 0\overset{m}{.}09$ for $A_B = 1\overset{m}{.}0$, hence the decrease in number counts as a function of extinction is considerably slower. Figure 1 shows the predicted surface number density of galaxies for DENIS and for $B < 19$, as a function of Galactic foreground extinction.

The NIR becomes notably more efficient at $A_B \simeq 2 - 3^m$. At an extinction of $A_B \simeq 3 - 4^m$, J becomes superior to I_c, while at $A_B \simeq 10^m$, K_s becomes superior to J. We can expect to find galaxies in J and K_s, even at $A_B = 10^m$. These are very rough predictions, do not take into account the most recent expected DENIS high-latitude limits (Mamon *et al.*, 1998), which are more favourable for the I and J bands, but less favourable for the K band (see § 1). Moreover, we have not yet taken into account any dependence on morphological type, surface brightness, orientation and crowding, which may lower the counts of actually detectable galaxies counts (Mamon, 1994).

In April 1997, a new cooling system for the focal instrument of DENIS has been mounted. This appears to increase the K_s band limiting magnitude by ~ 0.5 magnitude and therewith the number of galaxies detectable in the deepest obscuration layer of the Milky Way by a factor of about 2. Consequently, the *long dashed curve* representing the K_s counts in Figure 1 should be moved up by roughly a factor of 2, which would make the K_s

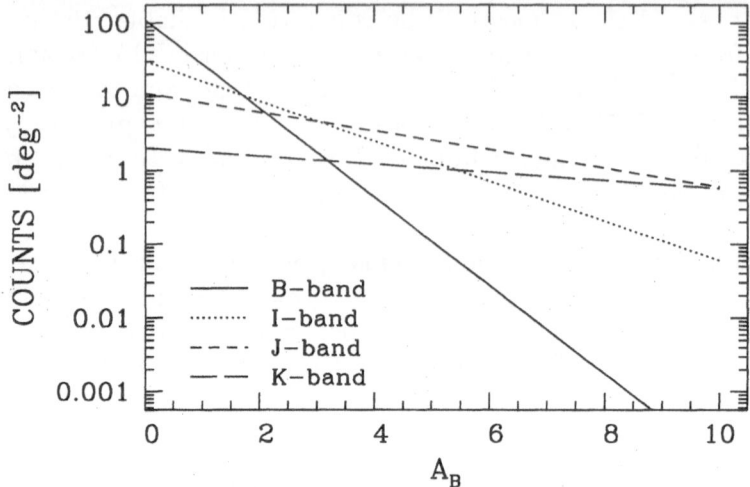

Figure 1. Predicted galaxy counts in B, I_c, J and K_s as a function of absorption in B, for highly complete and reliable DENIS galaxy samples and a $B_J \leq 19^m$ optical sample.

passband competitive with J starting at $A_B \simeq 7^m$.

3. DENIS data in the Norma cluster A3627

3.1. RECOVERY OF GALAXIES FOUND IN THE B BAND

Three high-quality DENIS strips cross the cluster Abell 3627 practically through its center. We inspected 66 images which cover about one-eighth of the cluster area within its Abell-radius of $R_A = 1°.75$ (each DENIS image is 12′x12′, offset by 10′ in declination and right ascension). The extinction over the regarded cluster area varies as $1^m.2 \leq A_B \leq 2^m.0$.

We cross-identified the galaxies found in the optical survey with the DENIS I_c, J, and K_s images. An example of a DENIS image in the central part of the cluster is given in Figure 3 of Paper I. On the 66 images, 151 galaxies had been identified in the optical. We have recovered 122 galaxies in the I_c band, 100 in the J band, and 74 in the K_s band (not including galaxies visible on more than one image). As suggested by Figure 1, the K_s band indeed is not optimal for identifying obscured galaxies at these latitudes due to its shallow magnitude limit. Most of the galaxies not rediscovered in K_s are low surface brightness spiral galaxies.

Surprisingly, the J band provides better galaxy detection than the I_c band. In the latter, the severe star crowding makes identification of faint galaxies very difficult. At these extinction levels, the optical survey does remain the most efficient in *identifying* obscured galaxies.

3.2. PHOTOMETRY OF GALAXIES IN THE NORMA CLUSTER

We have used a preliminary galaxy pipeline (Mamon *et al.*, 1997b, 1998), based upon the SExtractor package (Bertin & Arnouts, 1996) on the DENIS data in the Norma cluster to obtain I_c, J and K_s Kron photometry. Although many of the galaxies have a considerable number of stars superimposed on their images, magnitudes derived from this fairly automated algorithm agree well with the few known, independent measurements.

Magnitudes could be determined for 109, 98 and 64 galaxies of the 122, 100, 74 galaxies re-discovered in I_c, J, and K_s. Figure 2 shows the luminosity function (LF) of these galaxies together with the B band LF of the 151 galaxies visible on the same 66 DENIS images. The histograms are normalised to the area covered by the 66 images. The hashed area marks the 60 galaxies common to all 4 passbands. This subsample is mainly restricted by the K_s band. The magnitudes in the bottom row are corrected for extinction. The corrections are derived from Mg_2-indices of elliptical galaxies in the cluster (Woudt *et al.* in prep.) and interpolations according to the Galactic H I distribution.

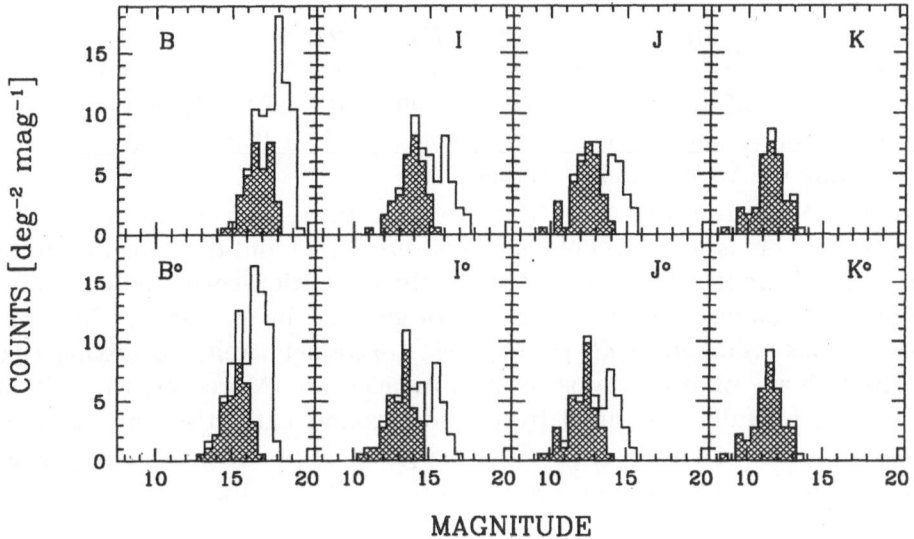

Figure 2.   The luminosity function for the observed Norma galaxies in B, I_c, J, and K_s. The bottom panels display magnitudes corrected for foreground extinction. The *hashed histograms* represent the sample common to all 4 passbands ($N = 60$).

To assess whether the LFs displayed here are, in fact, representative of the cluster as a whole — and therefore the extinction corrected NIR I_c, J, and K_s band LFs displayed in the lower panels characteristic for rich clusters — we compared the B band LF of the 151 galaxies on the 66 DENIS-images with the cluster LF as a whole (*cf.* Woudt, 1997). The extinction-corrected blue cluster LF of the 609 galaxies within the Abell radius, scaled to the Abell area, actually has lower number counts than the B^o band LF displayed in the bottom panel of Figure 2. This is explained by the fact that our three strips cross the center of the cluster and therewith the region of highest density. The comparison indicates that we are fairly complete to a magnitude of $B^o = 16\overset{m}{.}5$, which is more or less the shaded area, and that the shape of the total LF is very similar to the distribution of the common subsample.

Even though these LFs are still preliminary (we have so far covered only a small area of the Norma cluster and will have missed dwarf galaxies and other LSB galaxies due to the foreground obscuration) the here determined extinction-corrected LFs of the galaxies common to all passbands can be regarded as a first indication of the bright end of the NIR I_c, J, and K_s band LFs in rich clusters.

From the below discussed colours of the Norma galaxies, we know that the extinction corrections are of the correct order. Adopting a distance to A3627 of 93 Mpc (Kraan-Korteweg *et al.*, 1996), thus $m - M = 34\overset{m}{.}8$, the 60 galaxies cover a luminosity range in K_s of $-25\overset{m}{.}3 < M_K^o < -21\overset{m}{.}8$. This compares well with the bright end of the K_s band LF of the Coma cluster core derived by Mobasher & Trentham (1997), although it remains puzzling why the number counts derived by them (*cf.* their Table 1) are so much lower compared to the A3627 cluster.

The NIR magnitudes have been used to study the colour–colour diagram $I - J$ versus $J - K$. This has been presented and discussed in detail in Paper I. Here it suffices to state that the extinction-corrected colours of the cluster galaxies match the colours of galaxies in unobscured high latitude regions (Mamon *et al.* 1997b, 1998) extremely well, suggesting that our preliminary photometry is reasonably accurate. Moreover, the shift in colour can be fully explained by the foreground extinction or, more interestingly, the NIR colours of obscured galaxies provide, in principle, an independent way of mapping the extinction in the ZOA (see also Mamon *et al.*, 1997a).

4. 'Blind' search for galaxies

The GA is suspected to cross the Galactic Plane from the Norma cluster in the south towards the Centaurus cluster in the north. In this region, we performed a search for highly obscured galaxies on the so far existing DENIS survey images. The search area within the GA-region — marked as a *dashed box* in Figure 3 — is defined as $320° \leq \ell \leq 325°$ and $|b| \leq 5°$.

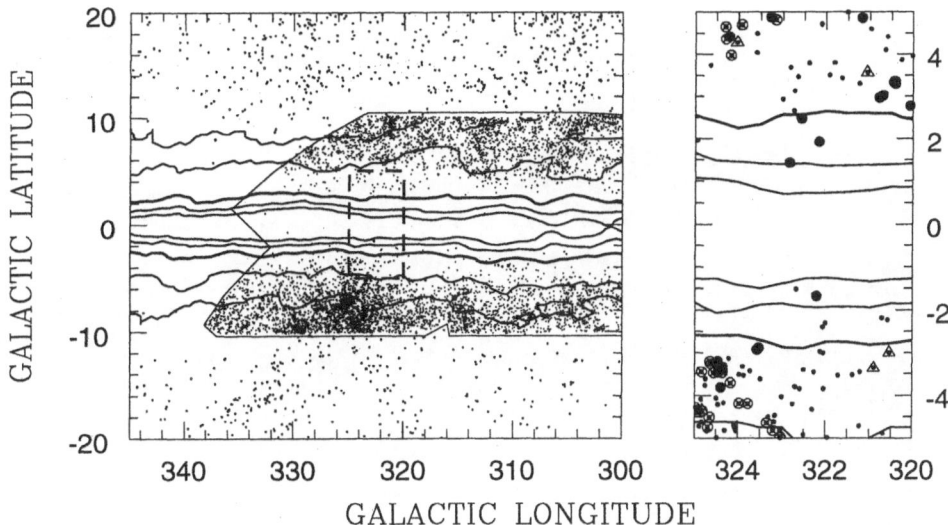

Figure 3.   Galaxy distribution in the GA region displaying Lauberts galaxies ($D \geq 1\rlap{.}'0$, Lauberts 1982) and galaxies from the deep optical search ($D \geq 0\rlap{.}'2$, outlined area). The superimposed *contours* represent absorption levels of $A_B = 1\rlap{.}^{m}5, 2\rlap{.}^{m}5, 5\rlap{.}^{m}0$ (*thick line*), $7\rlap{.}^{m}5$ and $10\rlap{.}^{m}0$, as determined from H I column densities and assuming a constant gas/dust ratio. The *box* marks the DENIS blind search area with the results shown enlarged in the right panel: optical galaxies re-identified on DENIS images ($N=31$, including 3 uncertain identifications) as *large encircled crosses*, optical galaxies not seen by DENIS ($N=6$) as *triangles*, and newly identified, optically invisible galaxies ($N=15$) as *filled dots*.

Of the 1800 images in this area we have inspected 385 by eye (308 in K_s). 37 galaxies at higher latitudes were known from the optical survey. 28 of these could be re-identified in the I_c band, 26 in the J band, and 14 in the K_s band. They are plotted as *encircled crosses* in Figure 3. In addition, we found 15 new galaxies in I_c and J, 11 of which also appear in the K_s band (*filled circles*). The ratios of galaxies found in I_c compared to B, and of K_s compared to I_c are higher than in the Norma cluster. This is due to the higher obscuration level (starting with $A_B \simeq 2\rlap{.}^{m}3 - 3\rlap{.}^{m}1$ at the high-latitude border of the search area, *cf. contours* of Fig. 3).

On average, we have found about 3.5 galaxies per square degree in the I_c band. This roughly agrees with the predictions of Figure 1, although the number of the inspected images and detected galaxies are too low to allow a statistical conclusion. Since we looked in an overdense region we expect *a priori* more galaxies. On the other hand, we do not expect to find galaxies below latitudes of $b \simeq 1° - 2°$ in this longitude range (Mamon, 1994). The visual impression of the low-latitude images substantiates this — the images are nearly fully covered with stars.

Figure 4 shows a few characteristic examples of highly obscured galaxies

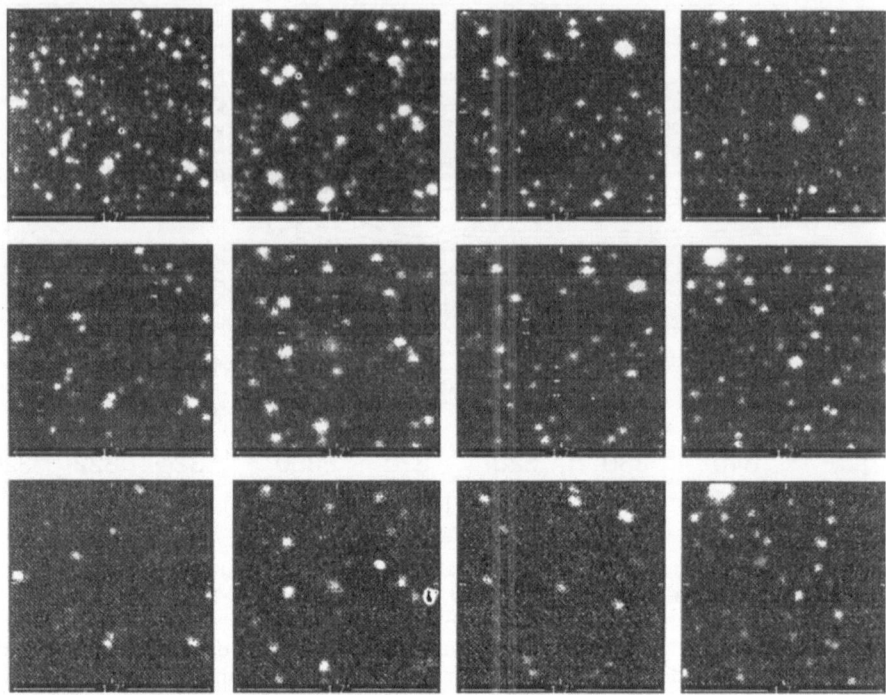

Figure 4.   DENIS survey images (before bad pixel filtering) of four galaxies found in the deepest extinction layer of the Milky Way; the I_c band image is at the *top*, J in the *middle* and K_s at the *bottom*.

found in the DENIS blind search. I_c band images are at the top, J in the middle and K_s at the bottom. The left-most galaxy is located at $(l, b) = (324°6, -4°5)$, with $A_B = 2^m8$ as estimated from H I-column densities (Kerr *et al.*, 1986) following the precepts of Burstein & Heiles (1982). It is barely visible in the J band, although its B band image is similar to the B of the second galaxy. This galaxy at $(l, b) = (324°7, -3°5)$ is, however, subject to heavier extinction $(A_B = 3^m7)$ and hence easier to recognise in the NIR. The most distinct image is the J band. The third galaxy at even higher extinction $(l, b, A_B) = (320°1, +2°5, 4^m6)$ is not visible anymore in the B band. Neither is the fourth galaxy: at $b = +1°9$ and $A_B = 6^m3$ this galaxy is not even visible in the I_c band and very faint in J and K_s.

The most important result from this search is that *highly obscured, optically invisible galaxies can indeed be unveiled in the NIR* and — as indicated with the distribution in the right panel of Figure 3 — found at lower latitudes than the deep optical survey. The lowest Galactic latitude at which we found a galaxy is $b \simeq 1.5°$ and $A_B \simeq 7^m5$.

5. Galaxies detected in H I

NIR surveys are the only tools that will identify early-type galaxies and therewith uncover the cores of massive groups and clusters at very low-latitudes. In addition, highly obscured spiral galaxies should be detectable with these surveys as well. Such identifications will proof important in connection with the systematic blind H I survey currently conducted with the Multibeam Receiver (13 beams in the focal plane array) at the 64 m Parkes telescope: a deep survey with a 5σ detection limit of 10 mJy is being performed in the most opaque region of the southern Milky Way ($213° \lesssim \ell \lesssim 33°$; $|b| \lesssim 5°$) for the velocity range of $-1000 \lesssim v \lesssim 12000$ km s^{-1}(Staveley-Smith, 1997). Roughly 3000 detections are predicted. Hardly any of them will have an optical counterpart. However, at these latitudes many might be visible in the NIR. The combination of data from these two surveys, *i.e.*, NIR photometry with H I-data (velocity and linewidth) will proof particularly interesting because it will allow the extension of peculiar velocity data *into* the ZOA via the NIR Tully–Fisher relation.

Only a few cross-identifications were possible with the data available from both surveys by June 1997. But we could identify thirteen galaxies detected blindly in H I on existing DENIS images. Four of them are visible in the B, I_c, J, and K_s bands. The other galaxies are only seen in the NIR. Four of them need further confirmation.

Figure 5 shows four examples of the candidates. The first galaxy is a nearby ($v = 1450$ km s^{-1}) ESO-Lauberts galaxy (L223-12) at $b = +4°8$ and $A_B = 3^m2$. It is very impressive in all three NIR passbands (note the larger image scale for this galaxy, *i.e.*, $3'3$ instead of $1'7$). The second galaxy at $(l, b, A_B) = (306°9, +3°6, 3^m3)$ is slightly more distant ($v = 2350$ km s^{-1}). This galaxy has also been identified in B and is quite distinct in I_c and J. The third galaxy at $(b, A_B) \simeq (-2°9, 4^m6)$ had been detected by us as an OFF-signal at $v = 2900$ km s^{-1} during pointed H I observations in the ZOA. It has no optical counterpart but can be clearly seen in all three NIR passbands. The last example is an uncertain NIR counterpart at $(b, A_B) \simeq (+1°5, 7^m5)$ of a galaxy detected in H I at $v = 1450$ km s^{-1}. It is barely visible in the I_c band.

Although the present data is scarce, NIR counterparts of H I detected, highly obscured galaxies certainly seem to merit a systematic exploitation for large-scale structure investigations.

6. Conclusion

Our pilot study illustrates the promises of using the NIR surveys for extra-galactic large-scale studies behind the ZOA as well as for the mapping of the Galactic extinction.

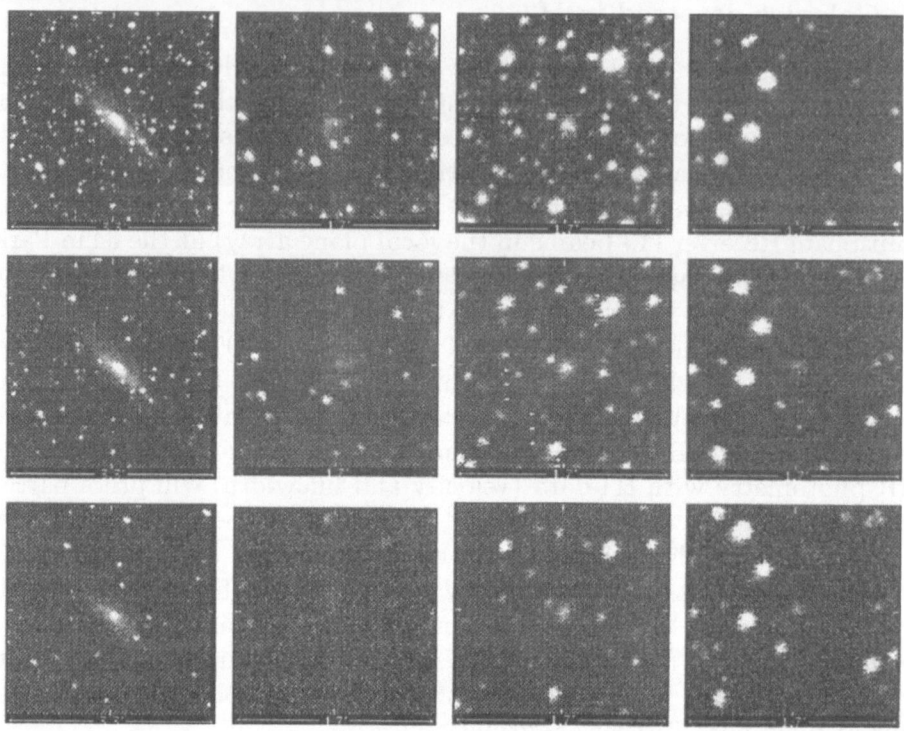

Figure 5.   DENIS survey images (before bad pixel filtering) of four galaxies detected blindly in H I at $|b| \leq 5°$; the I_c band image is at the *top* , J in the *middle* and K_s at the *bottom*.

At intermediate latitudes and extinction ($5° < |b| < 10°$, $1^m \lesssim A_B \lesssim 4 - 5^m$) optical surveys remain superior for identifying galaxies. However, the NIR luminosities and colours together with extinction data from the NIR colours will prove invaluable in analysing the optical survey data and their distribution in redshift space, and in the final merging of these data with existing sky surveys. Despite the high extinction and the star crowding at these latitudes, I_c, J and K_s photometry from the survey data can be successfully performed at these low latitudes and lead, for instance, to the preliminary I_c^o, J^o and K_s^o galaxy luminosity functions in A3627.

At low latitudes and high extinction ($|b| < 5°$ and $A_B \gtrsim 4 - 5^m$) the search for 'invisible' obscured galaxies on existing DENIS-images implicate that NIR-surveys can trace galaxies down to about $|b| \simeq 1°\!.5$. The J band was found to be optimal for identifying galaxies up to $A_B \simeq 7^m$, although this might change in favour of K_s with the new cooling system. NIR surveys can hence further reduce the width of the ZOA. This is furthermore the only tool that permits the mapping of early-type galaxies — tracers of density peaks — at high extinction.

The combination of two different surveys, *i.e.*, NIR data for highly obscured spiral galaxies detected in a systematic blind H I survey — a fair fraction could indeed be re-identified on DENIS-images — allows the mapping of the peculiar velocity field in the ZOA through the NIR Tully–Fisher relation. This will be pursued as well at intermediate latitudes ($5° < |b| < 10°$) with pointed H I observations of optically identified spiral galaxies. About 300 spiral galaxies have already been detected (Kraan-Korteweg *et al.* 1997).

Whether the systematic identification of ZOA galaxies from the DENIS survey must be performed by visual examination or whether galaxies can be successfully extracted using classical algorithms (Mamon *et al.* 1997b, 1998) or artificial neural networks (Bertin & Arnouts 1996, Bertin, in these proceedings) or a combination of both requires further exploration.

Acknowledgements

We thank Jean Borsenberger for providing bias subtracted, flat fielded DENIS images, Emmanuel Bertin for supplying recent updates of his SExtractor software package, and Eric Copet for providing software to display Figures 4 and 5.

References

Bertin, E., Arnouts, S., 1996, *Astr. Astrophys. Suppl. Ser.* **117**, 398
Burstein D., Heiles C., 1982, *Astron. J.* **87**, 1165
Cardelli J.A., Clayton G.C., Mathis J.S., 1989, *Astrophys. J.* **345**, 245
Epchtein, N., 1997, in *The Impact of Large Scale Near-Infrared Surveys* p. 15, eds. F. Garzon, N. Epchtein, A. Omont, W.B. Burton, B. Persi, Kluwer, Dordrecht
Epchtein, N., *et al.*, 1997, *Messenger*, **87**, 27
Gardner, J.P., Sharples, R.M., Carrasco, B.E., Frenk, C.S., 1996, *MNRAS* **282**, L1
Kerr, F.J., Bowers, P.F., Jackson, P.D., Kerr, M., 1986, *Astr. Astrophys. Suppl. Ser.* **66**, 373
Kraan-Korteweg, R.C., Woudt, P.A., 1994, in *Unveiling Large-Scale Structures Behind the Milky Way*, p. 89, eds. C. Balkowski, R.C. Kraan-Korteweg, ASP Conf. Ser. 67
Kraan-Korteweg R.C., Loan A.J., Burton W.B., Lahav O., Ferguson H.C., Henning P.A., Lynden-Bell D., 1994, *Nature* **372**, 77
Kraan-Korteweg, R.C., Fairall, A.P., Balkowski, C., 1995, *Astr. Astrophys.* **297**, 617
Kraan-Korteweg R.C., Woudt P.A., Cayatte V., Fairall A.P., Balkowski C., Henning P.A., 1996, *Nature* **379**, 519
Kraan-Korteweg, R.C., Woudt, P.A., Henning, P.A., 1997, *PASA* **14**, 15
Lauberts, A. 1982, The ESO/Uppsala Survey of the ESO (B) Atlas, ESO, Garching
Mamon G.A., 1994, in *Unveiling Large-Scale Structures Behind the Milky Way*, p. 53, eds. C. Balkowski, R.C. Kraan-Korteweg, ASP Conf. Ser. 67 (astro-ph/9405056)
Mamon, G.A., Banchet, V., Tricottet, M. Katz, D., 1997a, in *The Impact of Large-Scale Near-Infrared Surveys*, p. 239, eds. F. Garzon, N. Epchtein, A. Omont, W.B. Burton, B. Persi, Kluwer, Dordrecht (astro-ph/9608077)

Mamon, G.A., Tricottet, M., Bonin, W., Banchet, V., 1997b, in XVIIth Moriond Astrophysics Meeting on *Extragalactic Astronomy in the Infrared*, p. 369, eds. G. A. Mamon, Trinh Xuân Thuân, and J. Trân Thanh Vân, Frontières, Paris (astro-ph/9711281)

Mamon, G.A., Borsenberger, J., Tricottet, M., Banchet, V., 1998, in these proceedings

Mobasher, B., Trentham, N., 1997, *MNRAS*, in press (astro-ph/9708226)

Schröder, A., Kraan-Korteweg, R.C., Mamon, G.A. Ruphy, S., 1997, in XVIIth Moriond Astrophysics Meeting on *Extragalactic Astronomy in the Infrared*, p. 381, eds. G. A. Mamon, Trinh Xuân Thuân, and J. Trân Thanh Vân, Frontières, Paris, (paper I, astro-ph/9706093)

Staveley-Smith, L., 1997, *PASA* **14**, 111

Strutskie, M.F., *et al.* 1997, in *The Impact of Large Scale Near-Infrared Surveys*, p. 25, eds. F. Garzon, N. Epchtein, A. Omont, W.B. Burton, B. Persi, Kluwer, Dordrecht

Woudt, P.A., 1997, *Ph.D. thesis, Univ. of Cape Town.*

Woudt, P.A., Kraan-Korteweg, R.C., Fairall, A.P., Böhringer, H., Cayatte, V., and Glass, I.S., 1997, *Astr. Astrophys.*, in press

RAPID EXTRACTION OF GALAXIES FROM DENIS

G. PATUREL, I. VAUGLIN, M.C. MARTHINET AND C. PETIT

CRAL-Observatoire de Lyon
69561 Saint-Genis Laval, France
patu@obs.univ-lyon1.fr

AND

J. BORSENBERGER, L. PROVOST

Institut d'Astrophysique de Paris, France

1. Introduction

In order to face the challenge of collecting data for deep extragalactic samples it is necessary to identify new galaxies and to collect the main information on them: coordinates, magnitudes, diameters and axis ratios, position angles and morphological types. The DENIS project is a great opportunity to do that. It is the first time that such a survey is undertaken by getting CCD images directly at the telescope for an entire hemisphere. This new way of making a survey provides us with incredibly high quality images compared with photographic counterparts.

The Lyon-Meudon Extragalactic database (LEDA) was built in 1983 with the aim of collecting the main astrophysical parameters for the principal galaxies. Presently, LEDA contains more than 140,000 galaxies for which we collected the most important astrophysical parameters (Paturel et al., 1997). Our task in the DENIS consortium is to perform cross-identifications between DENIS and LEDA galaxies. This will allow us to recognize new galaxies for which it is important to follow kinematics up.

We present here the reduction of the first part of this quick look on the DENIS Survey. The raw data will be available through LEDA, while the provisional catalogue of mean data will be distributed to the DENIS consortium.

N. Epchtein (ed.),
The Impact of Near-Infrared Sky Surveys on Galactic and Extragalactic Astronomy, 221-227.
© 1998 *Kluwer Academic Publishers.*

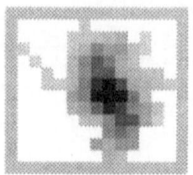

Figure 1.   A matrix of pixels for a new galaxy identified in I-band

2. Quick Data Reduction

2.1. PIPE-LINE OF DATA

All raw images send from Chile to France by P. Fouqué are pre-processed at the Institut d'Astrophysique de Paris by J. Borsenberger for cleaning purposes. The routines warn us when an image is ready to be analyzed. Then, each I-image is reduced (averaged 2x2 pixels), thresholded, compressed and send to Lyon by ftp. Each night about 3600 I-band images are received via this pipe-line. This flow is daily controlled in Lyon by C. Petit and J. Rousseau. The source extraction may start.

2.2. SOURCE EXTRACTION AND GALAXY RECOGNITION

The main method for source extraction is described in Paturel et al.(1996). Each object (star or galaxy) is represented by a matrix of pixels (Fig.1) from which a set of parameters is extracted: surface, total flux, peak flux, axis ratio, diameter, etc...

These parameters are used in a conventional discriminant analysis in order to perform a separation between stars and galaxies. The program was trained on a sample of 511 objects (239 stars and 272 galaxies) for which the discrimination between stars and galaxies was made by an human expert (F. Gallet and R. Garnier). The ratio of $maximum - intensity/surface$ appears as one of the most efficient parameter for such a discrimination. The principal component allows the recognition of 99.2% of objects from a test sample. Two levels of probability are chosen. The range between these levels define the uncertainty zone (see Fig.2 for illustration).

Both levels are chosen in such a way that 0% of stars will be identified as galaxy and that 10% of galaxies may be identified as stars. This choice will avoid the contamination of LEDA by stars.

The astrometry is made using the GSC as described in Paturel et al. (1996) and/or using the telescope coordinates registered in the header of each image. The accuracy is generally better than 10 arcsec.

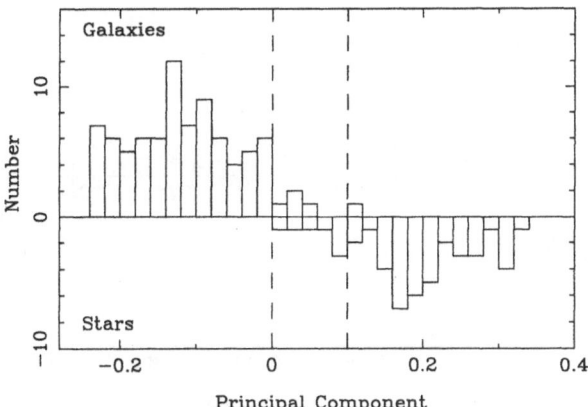

Figure 2. Galaxy/star recognition from a discriminant analysis. Histogram of the principal component obtained for our test sample. The two vertical lines define the uncertainty zone.

The final step of this treatment consists in checking visually all frames recognized as galaxies. This tedious part allows us to reject artefacts (a few percent) like those produced by star halos truncated by the edge of the frame. Such tuncated halos look like an elongated, low-surface brightness object, easily accepted as galaxy. This cleaning part is made by J.Rousseau and R. Garnier.

2.3. PROVISIONAL CALIBRATION

Flux are calibrated by comparison with the measurement in I-band photometry made by Mathewson et al. (1992, 1996). This comparison allows us to correct for seasonal variation of the zero-point. Fig.3 shows such a variation. After correction the zero-point distribution is quite Gaussian with an standard deviation of 0.20 magnitude (Fig.4).

In Fig.5, the comparison between Mathewson et al. and DENIS I-band magnitudes is shown. The standard deviation is 0.18 magnitude. If we assume that the error is identical for both systems we conclude that the mean error on DENIS extragalactic I-band magnitude is about 0.13 magnitude.

3. Cross–Identification with LEDA

3.1. AUTO-CROSS-IDENTIFICATION

Our first archive includes 52800 individual measurements. Some galaxies are measured twice because they are in the overlapping zone of the images.

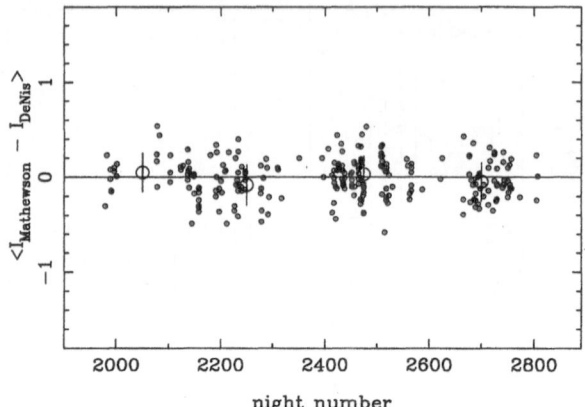

Figure 3. Zero-point variation obtained by comparison with Mathewson's I-band photometry.

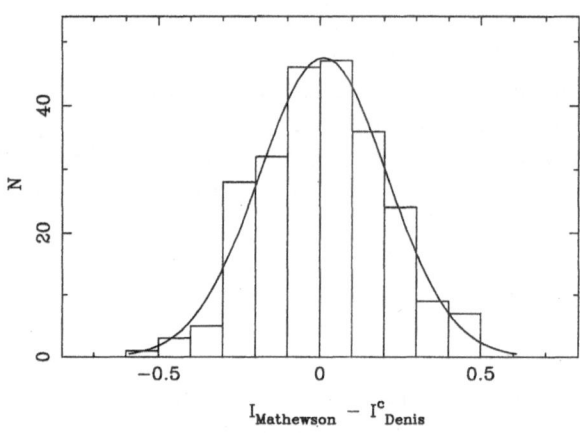

Figure 4. Zero-point distribution after a tiny seasonal correction

In the first step, we performe what we call the auto-cross-identification. This will consist in identifying measurements concerning a same object. This is done by a hierarchical method in which we merge step by step the closest measurements. The definition of the distance of two measurements i and j is taken as:

$$d_{ij} = \frac{1}{N} \sum_{k=1}^{N} \frac{|m_{ik} - m_{jk}|}{2\sigma_i}$$

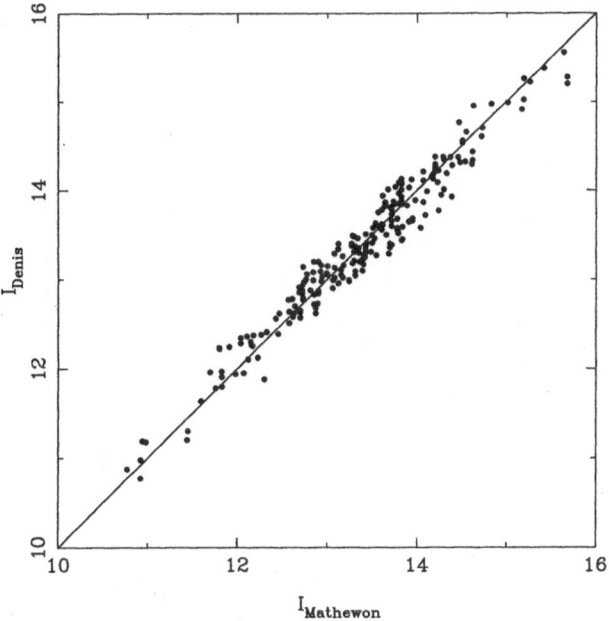

Figure 5. Comparison of extragalactic I-band magnitudes from Mathewson et al. and from DENIS

where, N is the number of parameters (coordinates, diameters magnitudes...) of a given measurement, m_{ik} is the k-th parameter of measurement i, m_{jk} is the k-th parameter of measurement j. σ_i is the standard error of the $i - th$ parameter. When d_{ij} is smaller than a given limit d_{limit} both measurements are merged and are replaced by a single one made of the mean of both. The final result does not depend on the order the original file is read. Note that, special care must be taken for periodic parameters, like right ascension and position angle (e.g., $p.a. = 0$ deg is identical to $p.a. = 180$ deg).

The distance d_{limit} is chosen from the distribution of all distances (Fig.6). We adopted $d_{limit} = 0.55$.

3.2. CROSS-IDENTIFICATION WITH LEDA GALAXIES

The previous step leads to the construction of a catalogue of mean parameters. Each galaxy of this catalogue must then be cross-identified with LEDA galaxies. This is done by calculating the distance between a DENIS galaxy (distance in the mathematical sense, as defined by the previous equation) and a LEDA galaxy. The cross-identification is performed if the distance is smaller than a limit d_{limit}. From a distribution of distances between LEDA

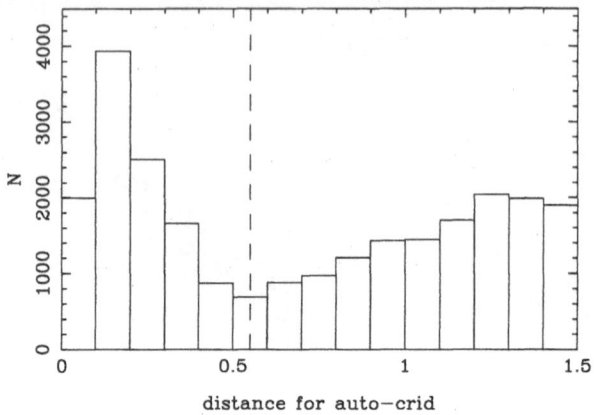

distance for auto−crid

Figure 6. Distribution of "distances" d_{ij} between two DENIS measurements (i and j). This graph is used to chose d_{limit} in the auto-cross-identification phase (see text).

and DENIS measurements similar with the one shown in Fig.6, we adopted $d_{limit} = 1.00$. This limit is voluntarily severe (i.e., small) because we prefer to miss a cross-identification than to merge two distinct galaxies. Finally, we rejected every galaxy which could be cross-identified with more than one LEDA galaxy.

3.3. DATA DISTRIBUTION

The raw data are available through LEDA database using either a telnet access or the World Wide Web.

```
telnet leda.univ-lyon1.fr
login: leda    (no password needed)
```

or

```
http://www-obs.univ-lyon1.fr/leda/leda-consult.html
```

This way allows the user to retrieve I-band raw parameters (including the matrix) for galaxies observed during the first year of the survey. These galaxies are identified with the provisional acronym RED for "Rapide Extraction DENIS" (e.g., RED1000). According to the "Data Release Policy Group" a code is required to access DENIS parameters. This code may be asked to denis@obs.univ-lyon1.fr.

On the other hand, the whole catalogue of 36246 galaxies (among which 21956 are new ones) is now available with all I-band mean parameters. This catalogue is available for DENIS Co-Investigators on request. To get it, send a mail to denis@obs.univ-lyon1.fr.

4. Follow–up

A galaxy without redshift is not very useful. Fortunately, for many DENIS galaxies identified with LEDA galaxies the redshift is already available (in LEDA more than 120.000 redshifts are collected). However, for the 21956 new galaxies detected after one year of the DENIS project, it is urgent to get at least the redshift and more detailed information if possible. In fact, this is the justification of this "Rapide Extraction". Thanks to this catalogue we may start redshift and HI follow-up's of the DENIS survey.

5. Acknowledgments

This work would have been impossible without technical help by F. GAL-LET, R. GARNIER, and J. ROUSSEAU from the Observatoire de Lyon. They are fully associated with this work.
The DENIS team is warmly thanked for making this work possible and in particular the operations team at La Silla headed by P. FOUQUE.

References

Paturel G., Garnier R., Petit C., Marthinet M.C. (1996) *A. & A.*, **311**,12
Paturel G., Andernach H., Bottinelli L., et al. (1997) *A.& A.* **124**,109
Mathewson D.S., Ford V.L., Buchhorn M. (1992) *Astrophys. J. Suppl. Ser.*, **81**,413
Mathewson D.S., Ford V.L. (1996) *Astrophys. J. Suppl. Ser.*, **107**,97

WIDE-ANGLE NEAR-INFRARED SURVEYS OF LOCAL GALAXIES

BAHRAM MOBASHER
Astrophysics Group,
Blackett Laboratory, Imperial College, Prince Consort Road,
London SW7 2BZ, U.K.

1. Introduction

Recently, it has become possible to perform wide-angle galaxy surveys at near-infrared ($2.2\mu m$) wavelengths. Such surveys cover limited areas to magnitude limits similar or fainter than that achieved by the DENIS or 2MASS projects. Therefore, the resulting number-magnitude counts can be used to calibrate these datasets. Also, due to the relatively bright magnitude limit of these wide-angle surveys, spectroscopic observations are straightforward and hence, one could construct the local near-ir luminosity function (LF) of galaxies. This could be used to predict the space density of galaxies expected from DENIS or 2MASS, providing reliable control samples for these larger studies.

The infrared LF is also required to model the counts of galaxies. This is particularly useful since the near-ir ($2.2~\mu m$) light is less affected by the evolutionary processes (due to the dominance of the near-infrared light by old, near-solar mass stars) and can be more securely modeled. Also, at these wavelengths the galaxy spectral energy distributions are similar for all galaxy types, leading to similar K-corrections for all galaxies. Therefore, the near-infrared number-magnitude counts of galaxies are less affected by the mix of galaxy types or uncertainties due to evolution, compared to other wavelengths.

In this article, I present a brief review of the available infrared surveys which are most comparable to those conducted by the DENIS and 2MASS teams. Using the recent wide-angle surveys, the LF of local galaxies at this wavelength will then be constructed.

N. Epchtein (ed.),
The Impact of Near-Infrared Sky Surveys on Galactic and Extragalactic Astronomy, 229-237.
© 1998 *Kluwer Academic Publishers.*

2. Near-infrared Surveys of Field Galaxies before DENIS and 2MASS

The main problem in performing infrared surveys is the small field of view of near-infrared detectors. Because of this, galaxy surveys at this wavelength require a trade-off between the area they cover and the depth of the survey. Therefore, previous near-ir galaxy surveys are either constructed from optically selected samples, covering relatively larger areas (Mobasher et al. 1993) or, are mainly pencil beam surveys (Glazebrook et al 1995; Cowie et al 1996), covering small solid angles to deeper levels. However, the former surveys suffer from optical selection criteria, making it difficult to convert to an infrared limited survey because of changes in the type mix with redshift whereas, the latter are affected by uncertainties due to small number statistics of galaxies of a given luminosity.

With the advent of large format near-ir detectors, wide-angle surveys at this wavelength have become possible. These provide the closest available surveys to DENIS and 2MASS (i.e. wide-angle, relatively shallow). There are three independent such programs currently in progress, including, a survey complete to K=15 (Gardner et al 1997), a deeper survey, covering a smaller area and complete to K=16 (Szokoly et al 1998) and a larger survey (with not yet available spectroscopic measurements) to K=16 (Huang et al 1997).

Partial redshift information are available for both Gardner et al (1997) and Szokoly et al (1998) surveys. Work is currently in progress to complete the spectroscopic observations of Szokoly et al (1998) survey to K=16.5. Moreover, spectroscopic observations are planned to measure redshifts for galaxies in Huang et al (1997). The list of different wide-angle, near-infrared galaxy surveys (with available or planned spectroscopic data) is presented in Table 1. These provide the control samples to compare with the much larger DENIS and 2MASS projects.

However, there are a number of other surveys which extend much deeper over a smaller area (Cowie et al 1996; McLeod et al 1995; Djorgovski et al 1995; Moustakas et al 1997). For example, the survey by Cowie et al (1996) employs the LRIS spectrograph on the Keck to measure the spectra of galaxies to K=20 mag., using two of the Hawaii Deep Fields, to establish the near-ir LF of galaxies at high redshifts. These deep surveys are not comparable with the DENIS and 2MASS and hence, will not be further discussed here.

3. Near-infrared Field Luminosity Function

Near-infrared redshift surveys of galaxies, complete to a given magnitude limit, are needed to construct the luminosity function at this wavelength.

TABLE 1. The wide-angle infrared surveys currently available

Survey	Area	K_{lim}	Spect. completeness	Telescope (Detector)
Mobasher et al 1993	41.56 deg^2	12.25	98%	UKIRT (UKT9)
Glazebrook et al 1995	551.9 arcmin^2	16.5	59%	UKIRT (IRCAM)
Gardner et al 1997	4.4 deg^2	15	90%	1.3m KPNO (IRIM)
Szokoly et al 1997	0.8 deg^2	16	30%	1.3m KPNO (IRIM)
Huang et al 1997	9.8 deg^2	16	—	0.6m/2.2m UHT (NICMOS)

This is carried out using the surveys presented in Table 1. Over the years, different techniques have been developed to construct the LF of galaxies, correcting for various effects such as the incompleteness due to selection criteria (Mobasher et al 1993), inhomogeneities in galaxy distribution (Efstathiou et al 1988) and spectroscopic incompleteness (Sandage, Tamman and Yahil 1979).

In this section, a wide-angle near-ir survey, carried out in collaboration with my colleagues at Johns Hopkins University is presented. While the DENIS and 2MASS surveys will eventually constrain the bright-end of the local infrared LF, this survey is designed to establish its faint-end slope. A brief discussion of two different methods for constructing the LF of galaxies is carried out in sections 3.1.1 and 3.1.2. These will then be applied on this survey to estimate the local near-ir LF. A comparison between differnt measurements of the near-ir LF for field galaxies will then be carried out in section 4.

3.1. A WIDE-ANGLE MEDIUM DEEP NEAR-INFRARED SURVEY

The near-ir survey here is carried out using the NICMOS 3 (HgCdTe) 256×256 array (IRIM) on the 1.3m telescope at KPNO. Two fields with existing multi-colour photometric ($UB_J R_F I_N$) and spectroscopic data ($Lynx2$ and $SA68$) are observed to $K_s = 17$ mag (at 5σ detection). The filter used for these observations (K_s at $\lambda = 2.15\mu m$; $\Delta\lambda = 0.33\mu m$) is similar to that of the 2MASS. Details of the observation and data reduction are given in Szokoly et al (1998). The number counts from this survey is compared with other studies in Figure 1, showing agreement within errors. As the spectroscopic observations here are based on an optically selected sample, it is likely that at fainter K_s magnitudes, the survey becomes incomplete in redshift space. The incompleteness, defined as the change with magnitude of the ratio of the number of galaxies with measured redshift to the total number of galaxies to a given K_s limit, is corrected by assuming a Fermi-Dirac distribution for the selection function (Sandage, Tammann & Yahil (1979)), $\left(\exp\left(\frac{m-m_l}{\Delta m}\right)+1\right)^{-1}$. This parametric function is then fitted to the observed data in Figure 2.

We find $m_l = 15.25$, $\Delta m_l = 0.80$ for SA68 and $m_l = 14.86$, $\Delta m_l = 0.54$ for Lynx2. For the luminosity function analysis in this study, we select a redshift completeness limit of 30% in Fig. 2, corresponding to limiting magnitudes of $K_s = 16$ and $K_s = 15.25$ in the SA68 and Lynx2 fields respectively. There are a total of 110 galaxies brighter than these limits. The redshift distribution for the complete sample, containing both the fields, are presented in Figure 3. The redshift distributions show that most of the near-ir selected galaxies are local objects ($z < 0.4$) and reveal the presence of a cluster in the Lynx2 field at $z \sim 0.05$, which could affect estimates of the luminosity function. To explore the sensitivity of the LFs to density enhancement in the Lynx2 field, we employ two different methods for estimating the LF from this survey.

3.1.1. *The Traditional Method*

For each galaxy in the survey, the maximum comoving volume, V_{max}, within which it remains in the sample by satisfying its apparent magnitude and redshift limits is calculated as

$$V_{max} = (c/H_0)\Omega_i \int_{z_1}^{z_2} \frac{d_L^2\ dz}{(1+z)^3(1+2q_0z)^{0.5}}$$

where Ω_i is the solid angle of the field containing the galaxy and $z_1 = max(z_L, z_l)$ and $z_2 = min(z_U, z_u)$. z_L and z_U are, respectively, the lower and upper redshift limits of the survey and z_l and z_u correspond to the redshifts the galaxy would have when shifted, respectively, to the brighter and fainter apparent magnitude limits of the survey. The luminosity distance of the galaxy, d_L, is given as

$$d_L = (c/H_0q_0^2)(q_0z + (q_0 - 1)((1 + 2q_0z)^{0.5} - 1)))$$

The luminosity function is then estimated by summing contributions of galaxies in different absolute magnitude bins

$$\phi(M) = \Sigma_i(\frac{1}{V_{max}^i})$$

Applying this method on the survey in section 3.1, we construct the K-band LF and fit it to a Schecter form, using a χ^2 technique. The resulting Schecter LF parameters are given in Table 2 and shown in Figure 4. This LF is sensitive to the presence of density enhancements in the sample.

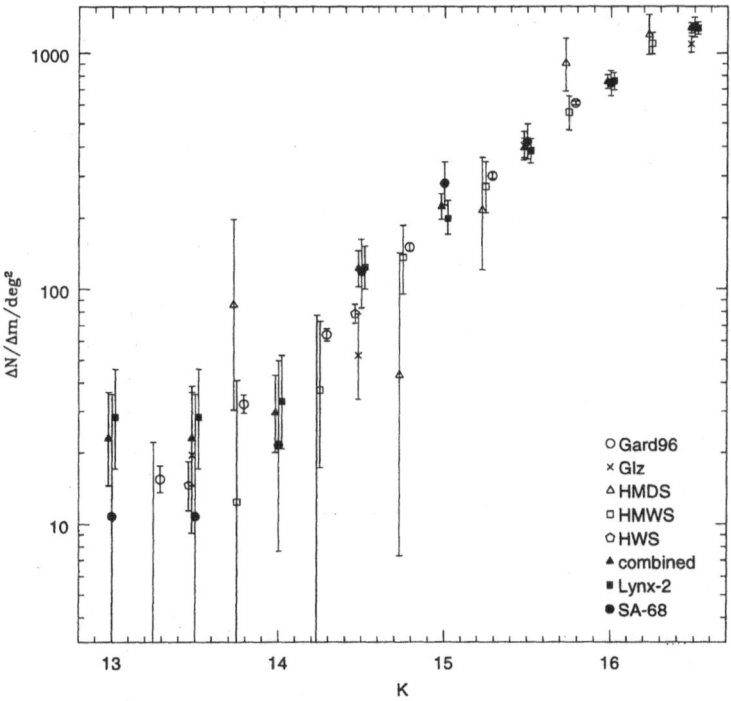

Figure 1. K-band differential number counts in the $K = 13 - 16.5$ range. The data are compiled by Gardner et al (1993). our SA-68 and Lynx-2 fields are also shown. Errors are 1σ estimates from the raw counts.

3.1.2. *The C-method*

The C-method is a non-parametric technique for constructing the LF of galaxies, insensitive to density inhomogeneities (Lynden-Bell 1971). This calculates the cumulative LF which is then differentiated to give the more familiar differential form. The cumulative LF, $\Psi(M_0)$, is defined as the density of galaxies with $M < M_0$. However, in reality we measure the cumulative distribution $X(M_0)$ which is the same as $\Psi(M_0)$ subject to a set of observational constraints (i.e. surface brightness limit). The C-method constructs the function $C(M)$, a subset of $X(M)$, for which the following relation satisfies:

$$\frac{d\Psi}{\Psi} = \frac{dX}{C}$$

In principle, $C(M_0)$ is the number of galaxies brighter than M_0 which would be observed if their absolute magnitude were M_0. Solving for $\Psi(M)$ leads

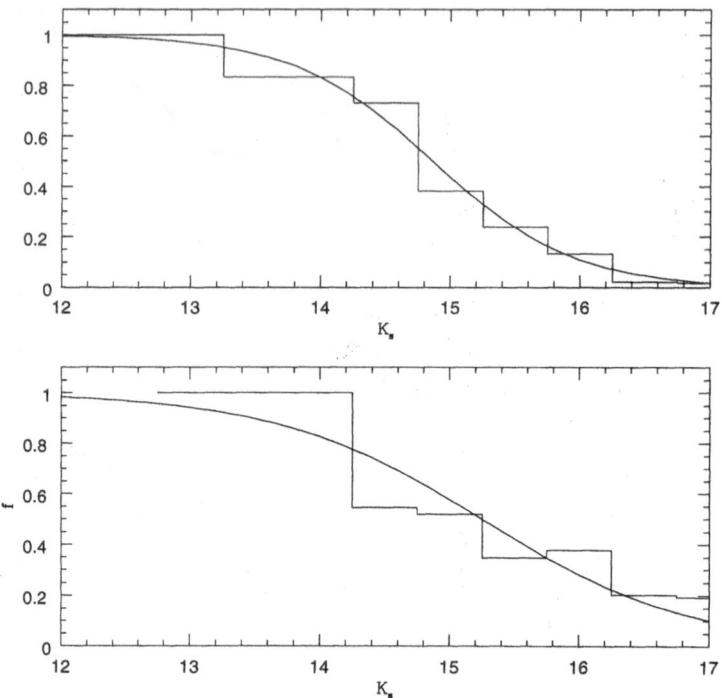

Figure 2. The redshift completeness, f, in each of the fields. The sample used in the luminosity function is cut at $K_s = 16$ in SA68 and $K_s = 15.25$ in Lynx2.

to

$$\Psi(M) = A \, exp \, (\int_{-\infty}^{M} \frac{dX}{C})$$

Differentiating this relation gives the differential LF.

The quantities dX, are series of Dirac Delta functions $dX = \Sigma_i \delta(M - M_i)$. Using the redshift information of galaxies in the sample, the functions $C(M)$ and $X(M)$ are calculated as explained in SubbaRao et al (1996). Knowing these functions, the differential LF can be calculated from the above relation.

Applying the C-method to our data, the near-ir LF is constructed and presented in Figure 4 (open circles). The errors are estimated using bootstrap re-sampling simulations. A parametric fit to the data, using the cluster-free maximum likelihood method gives the Schecter LF parameters, as listed in Table 2. This also shows that the shape of the infrared LF here is consistent with the Schecter form.

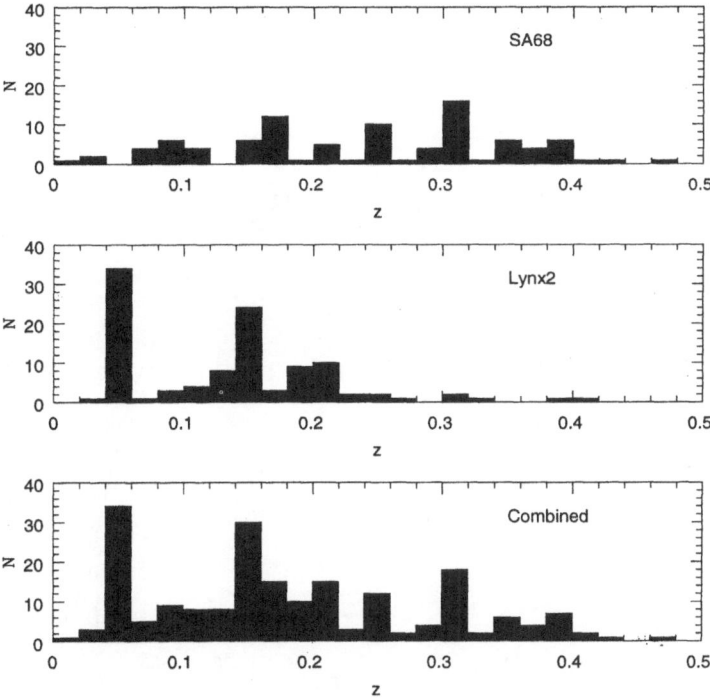

Figure 3.   The redshift distribution of each sample of galaxies that were used in the luminosity function derivation.

4. Comparison of the Infrared LFs

The near-ir LF parameters, estimated from the traditional and the C-method, are compared in Table 2. Also listed in this table are the *local* LFs from other independent studies, converted to the same scale as explained in Szokoly et al (1998). All the estimates here are corrected to $H_0 = 50$ Km/sec/Mpc and $q_0 = 0.5$. It is clear that, at a given α, the M_K^* values from different methods are in close agreement. However, the space density of local galaxies (ϕ^*) in the present sample is slightly smaller than others (~ 1.5).

Compared to a similar study by Gardner et al (1997), the LF found here has a steeper faint-end slope (Figure 4) while the two LFs are in close agreement at the bright-end. This result is currently being investigated by measuring redshifts for the fainter galaxies in our sample.

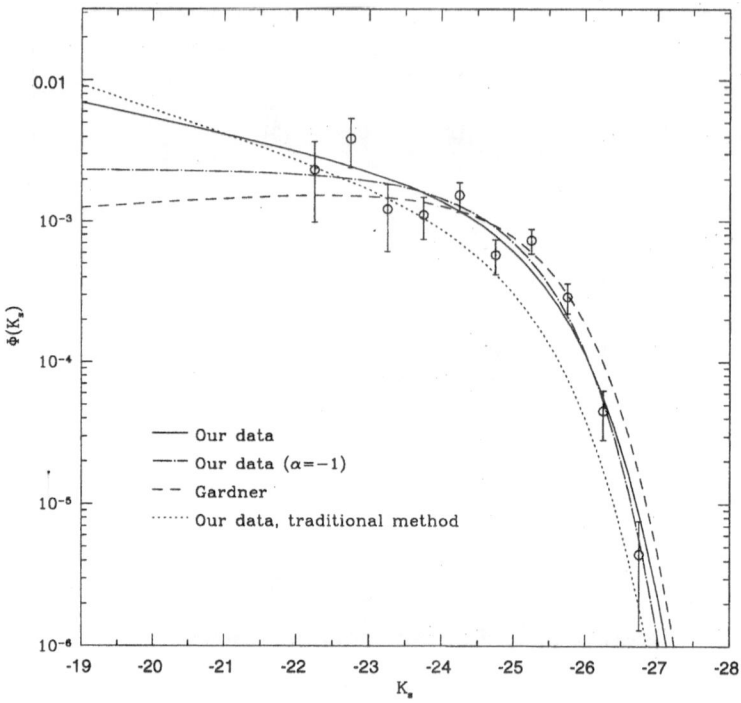

Figure 4. Differential luminosity function from the C-method (open circles) is compared with the fits to the LFs from the traditional method and Gardner et al (1997)

TABLE 2. Schecter infrared luminosity function parameters

Survey	M_K^*	α	Φ^*	n
This study (C-method)	−25.09	−1.27	0.15×10^{-2}	110
This study (conventional method)	−24.94	−1.42	0.95×10^{-3}	110
This study $\alpha = -1$	−24.84	−1	0.15×10^{-2}	110
Gardner et al 1997	−24.87	−1.03	0.22×10^{-2}	532
Glazebrook et al 1995	−24.55	−1.04	0.33×10^{-2}	98
Mobasher et al 1993	−24.88	−1	0.14×10^{-2}	95

5. Discussion

The local near-infrared surveys discussed in this article provide reliable control samples for both DENIS and 2MASS. Also, the near-ir LFs here provide estimates of the space density of galaxies in the local Universe expected from these surveys.

The steep faint-end slope of the local infrared LF, if confirmed, will have important implications towards constraining the models for formation of nearby, low-luminosity field galaxies (ie. the mergers scenarios). This is because the infrared LF is mainly sensitive to the mass function and not the star formation (i.e. young population) in galaxies. Also, this implies the existence of a large population of evolved, metal rich galaxies in the local Universe. The colour distribution of the K-selected surveys shows that the faint blue galaxies start to contribute to the galaxy counts at about $K \sim 18$ mag. (Gardner et al 1995). Such near-infrared surveys will then reveal if the faint blue galaxies have an underlying population of old stars. The surveys selected in the K-band are mainly dominated by normal massive galaxies. Therefore, the characteristic magnitude (M_K^*) found for the IRLF here implies that the very red galaxies with $M_K \sim -27$ mag. (Egami et al 1996) have a space density of $\leq 5 \times 10^{-5}$ Mpc^{-3} and hence, are not likely to be local objects ($z < 0.4$).

References

Cowie, L. L., Songaila, A., Hu, E. M., Cohen, J. G. 1996, Astron.J. 112, 839

Djorgovski, S. et al. 1995, Ap.J., 438, L13

Efstathiou, G.; Ellis, R. S.; Peterson, B. A. 1988, MNRAS, 232, 431

Egami, E.; Hu, E; Cowie, L.L 1996 Ap.J. 112, 73

Gardner, J. P., Sharples, R. M., Carrasco, B. E., Frenk, C. S. 1997, Ap.JL 480, L99

Glazebrook, K., Peacock, J. A., Collins, C. A. and Miller, L. 1994, MNRAS 266, 65

Huang, J. S., Cowie, L. L., Gardner, J. P., Hu, E. M., Songaila, A., Wainscoat, R. J., 1997, Ap.J. 476, 12

Lynden-Bell, D., 1971, MNRAS, 155, 95

Mobasher, B., Ellis, R. S. and Sharples, R. M. 1986, MNRAS 223, 11

McLeod, B.A., Bernstein, G.M. Reike,M.J., Tollestrup,E.V. & Fazio, G.G. 1995. Ap.JS 96, 117

Moustakas, L.,A., Davies, M., Graham,J.R., Silk,J., Peterson,B.A. & Yoshii, Y. 1997, Ap.J. 475, 44

Sandage, A., Tammann, G.A., & Yahil, A., 1979, Ap.J. 232, 352.

Saracco,P., Iovino,A.,Garilli,B., Maccagni,D. & Chincarini, G. 1997, AJ. 114, 887

SubbaRao, M.U., Connolly, A.J., Szalay, A.S. & Koo, D. C. AJ. 112, 929

Szokoly, G.P., Subbarao, M. U., Connolly, A.J. & Mobasher, B. 1998, Ap.J. 492 (in press)

DETECTION OF LOW SURFACE BRIGHTNESS GALAXIES IN 2MASS

T. JARRETT

IPAC — Caltech/JPL

Pasadena, CA 91125

Abstract. The study of low surface brightness galaxies (LSB) has become popular in recent years owing to their extreme properties (e.g., mass to light ratio) which make them attractive as the harbingers of the so called "missing" dark (baryonic) matter in the universe. The Two Micron All Sky Survey (2MASS) is expected to detect 1 million galaxies, a fraction of which will be LSB type galaxies. We have carried out a preliminary study of LSB galaxies using near-infrared data acquired with the 2MASS prototype camera. We report the results from a study of the Coma cluster and a field located near the north galactic pole.

1. Introduction

What is a low surface brightness galaxy (LSB)? Historically low surface brightness galaxies are delineated from their more "normal" cousins by their central surface brightness. In general, LSB galaxies have a central surface brightness fainter than 22 - 23 mag per sq. arcsec at B band (cf. Bothun et al. 1991). This loose definition allows a wide variety of objects classified as such, including spirals, ellipticals, dwarf ellipticals (probably the most numerous), dwarf irregulars and in the deepest studies, globular clusters (associated with massive galaxies). Because they are (possibly) abundant in number and their mass to light ratio is large, LSB galaxies may provide enough mass to cluster systems (e.g., Coma) to close or significantly narrow the "missing" dark matter gap. The true nature of these objects remains a mystery, including their stellar population(s), metallicity, star formation efficiency and evolution, and as noted, their overall mass. A number of large-area coverage studies have been carried out to address these issues (to cite just a few: Sandage et al. 1985; Schombert et al. 1992; Impey et al. 1996;

N. Epchtein (ed.),
The Impact of Near-Infrared Sky Surveys on Galactic and Extragalactic Astronomy, 239-246.

Bothun et al. 1993; Bernstein et al. 1995; Ulmer et al. 1996; Sprayberry et al. 1996; O'Neil et al. 1997; for a review, see Impey and Bothun 1997).

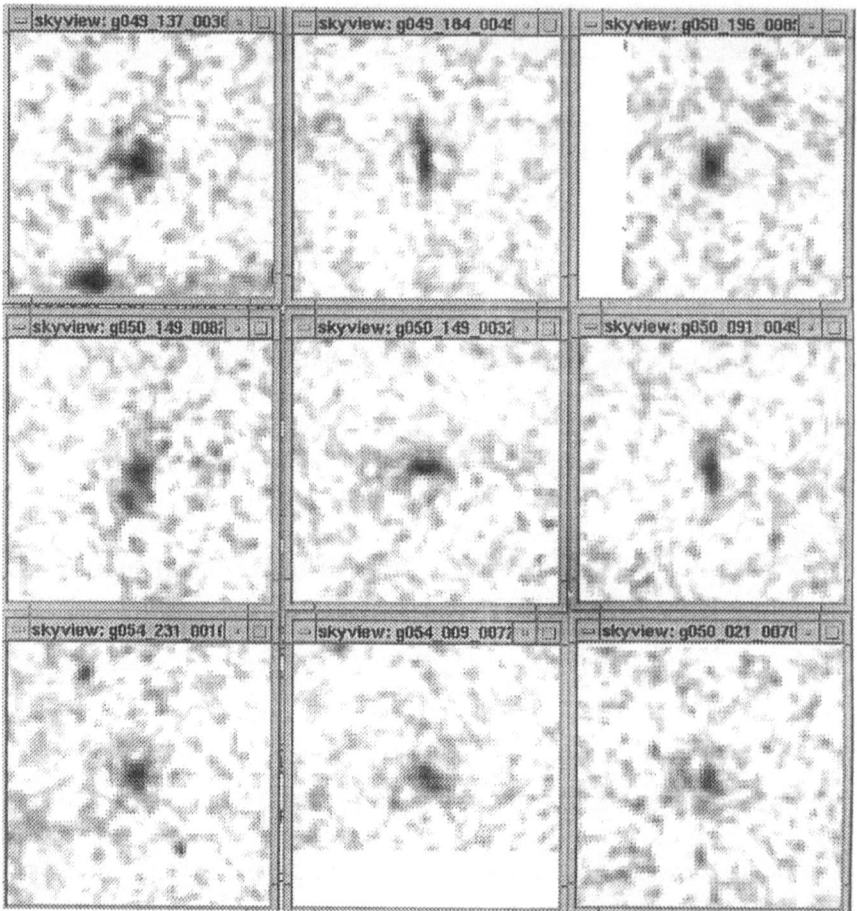

Figure 1. J-band "postage stamp" images of a representative sample LSB galaxies toward the Coma cluster. The images are about 1 arcmin in width. Integrated mags range from 14 to 15.

Adopting the central surface brightness criteria of 22 mag per sq. arcsec at B as the working definition of an LSB galaxy, we can estimate the surface brightness in the near-infrared using a color of (B-K) = 3 or 4 (B-I ranges from 2 to 3 for LSBs, O'Neil 1997, private comm). Thus, our definition of an LSB galaxy in the near infrared becomes

```
>19.0 - 20.0 mag per arcsec2 at J
>18.5 - 19.5 mag per arcsec2 at H
>18.0 - 19.0 mag per arcsec2 at K
```

The 2MASS survey will cover the entire sky, north and south, at the near-infrared bands of J, H and K, with a limiting J magnitude of about 15.0 for galaxies. 2MASS sacrifices depth for areal coverage, but still should be sensitive to the bright end of the LSB spectrum. For example, the typical sky noise in a 2MASS coadd image is:

```
J:
        1-sigma = 21.2 mag per sq. arcsec
        2-sigma = 20.5 mag per sq. arcsec
K:
        1-sigma = 19.9 mag per sq. arcsec
        2-sigma = 19.2 mag per sq. arcsec
```

Based on the flux limits of the survey and the typical background noise, 2MASS should be sensitive to LSBs with central surface brightness up to 20 mag per sq. arcsec, or equivalent to a B surface brightness of 23 to 24 mag per sq. arcsec. This corresponds to "bright" LSB galaxies, easily detected in deep photographic and CCD optical limited-coverage surveys now underway, but far from the state of the art (e.g., Rlim ∼27, O'Neil et al. 1997). Nevertheless, 2MASS should detect a wealth of LSB type galaxies due to the all sky coverage. This paper provides some preliminary information on what we should expect to see in 2MASS with LSB galaxies using data acquired for the Coma cluster (z=0.023), and the galactic pole region SA 57 .

Questions we address: How many LSB galaxies do we detect? Can these galaxies be well fit by an exponential (thus suggesting a population type)? How do their colors compare with normal galaxies? Outstanding Issues: Are normal or high surface brightness galaxies (HSB) and LSB galaxies distinct populations? Or, are LSB galaxies that 2MASS is likely to detect simply a "faint" extension of the HSB – normal – galaxy population? The latter point does not preclude the importance of LSB galaxies, it simply means that 2MASS may not have the sensitivity to address outstanding issues with regard to LSBs.

2. Data

The data was acquired with the 2MASS prototype camera (single channel mode), during observing runs at KPNO during spring of 1995. Five square degrees, comprising the SA57 region (NGP) and one square degree, centered on the Coma cluster, were examined in detail.

JHK 2MASS scans cover about 1 sq. degree (in 6 degree strips). One scan through Coma contains over 150 galaxies down to J∼15.

The sensitivity and depth of the 2MASS image product is demonstrated in Figure 1, in which we show some J-band examples of LSB galaxies found in the Coma cluster. The integrated J mag for the set ranges from 14 to 15.

3. Central Surface Brightness and Color

The central surface brightness was computed from the center or peak pixel value of the galaxy (for 2MASS, the pixel size is 1 arcsec, but the true resolution is closer to 2 arcsec). Figure 2 shows the central surface brightness vs. integrated mag for LSB galaxies, normal (or HSB) galaxies and for stars. The horizontal dashed lines denote the limit for LSB candidacy, as given by the adopted definition given in section 1. Note that LSB galaxies need only satisfy the limit in at least one band (thus, for example, a galaxy may satisfy the limit in K band, but not in J band).

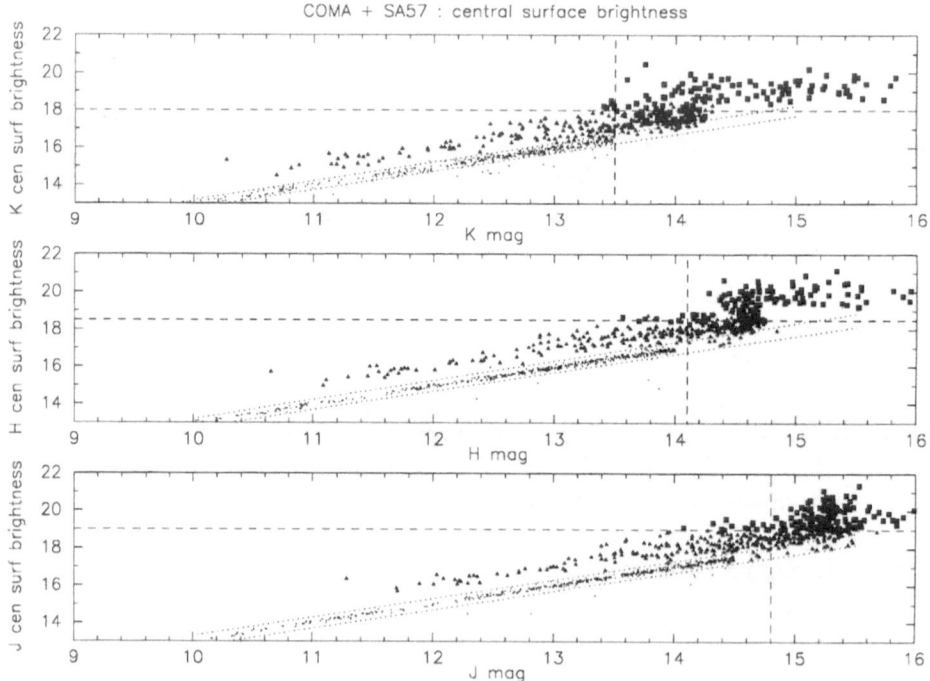

Figure 2.   Central surface brightness vs. the integrated flux. LSB galaxies are denoted by the filled squares, HSB galaxies by small triangles and stars by small points. The horizontal dashed lines represent the defined limit for LSBs – they need only satisfy the limit in at least one band. The vertical dashed lines represent the sensitivity (completeness) limit of the 2MASS survey.

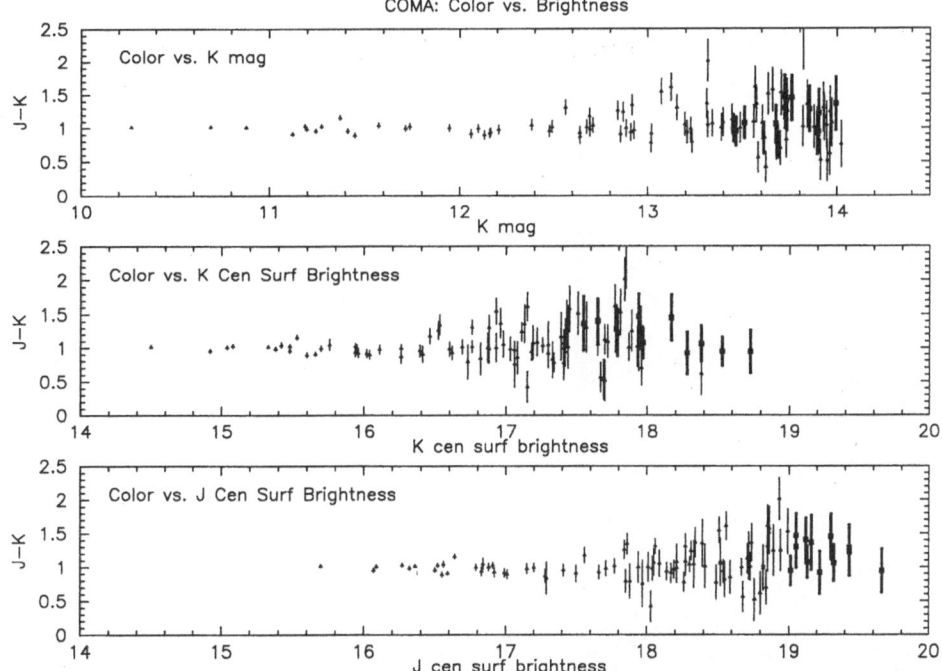

Figure 3. Color vs. brightness for Coma galaxies: LSBs are denoted by filled square symbols (and thicker error bars) and HSBs by small triangles.

There are a number of LSBs that are situated well above the HSB population, but they tend to be very faint (low SNR) subject to significant distortion from background noise. For the brighter LSB galaxies, the central surface brightness is not significantly different than HSB galaxies with comparable integrated flux (see J band results in particular).

Do LSB galaxies detected in 2MASS form a distinct population? It appears more likely that the Coma+Sa57 LSBs are simply an extension of the HSB or normal galaxy surface brightness curve. Moreover, the colors of LSB galaxies (see Figure 3) are typically about 1.0 in J-K, similar to HSB galaxies – particularly if you take into account the scatter in the color measurement.

We find that the peak intensity over

4. Radial Surface Brightness

An exponential function was fit to the mean surface brightness profile (per band) for each LSB galaxy. For the brighter galaxies in which the elliptical parameters are well determined, the radial profile was constructed from elliptical annuli. For most of the remaining LSB galaxies, the annuli were

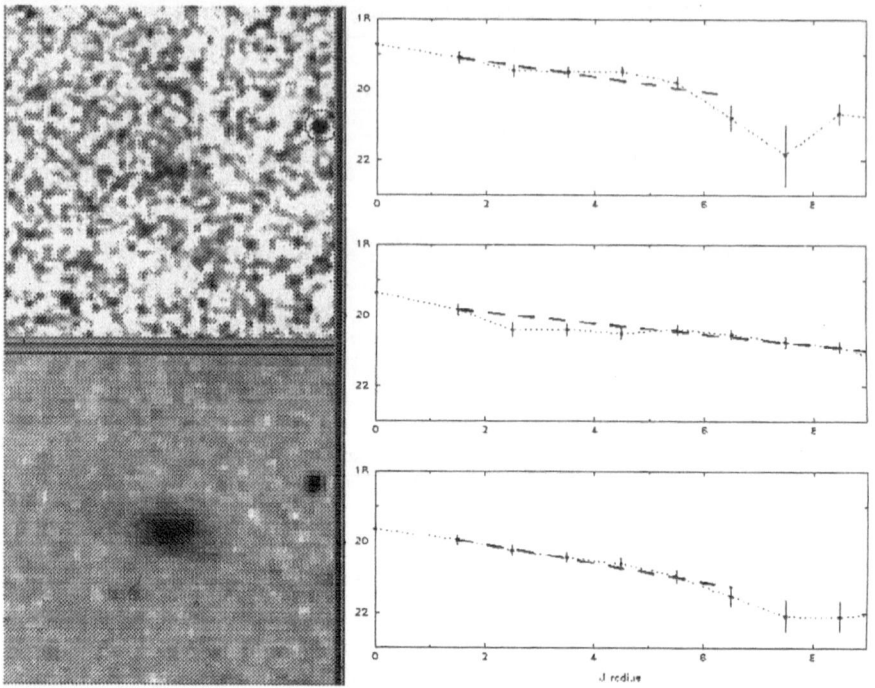

Figure 4. (a) SA57 galaxy as seen in K (upper panel), and optical DSS (lower panel). The integrated J mag is ∼14.9. (b) Radial surface brightness. K (top panel), H (middle) and J (lower panel). The dashed lines represent the best fit exponential. The derived scale length is ∼5 arcsec.

circular.

The exponential was fit over a range of 2 to 10 arc seconds – comprising most (if not) all of the area that is not lost in the background noise for a typical LSB galaxy. We avoid the inner 2 arc seconds to minimize the effect of the PSF. Only points with >2*sigma values were used in the fit, where sigma represents the uncertainty in the mean surface brightness measurement (poisson statistics). Figure 4 shows a typical LSB galaxy and its radial profile.

The derived scale length for Coma galaxies, Figure 5, ranged from a low of ∼1 arcsec (typical of the fainter LSBs – indicative of our lack of sensitivity), corresponding to about 500 pc (assuming H0 = 75 km/s/Mpc and DM = 34.9 for Coma) to 10 arcsec for bright LSBs with a very flat profile, corresponding to ∼4.6 kpc for Coma galaxies. No redshift data is available for the SA57 galaxies so the distance is unknown. Based on their scale lengths, however, the brighter SA57 galaxies are probably closer than the Coma cluster.

A simple exponential appears to fit the profiles very well in most cases.

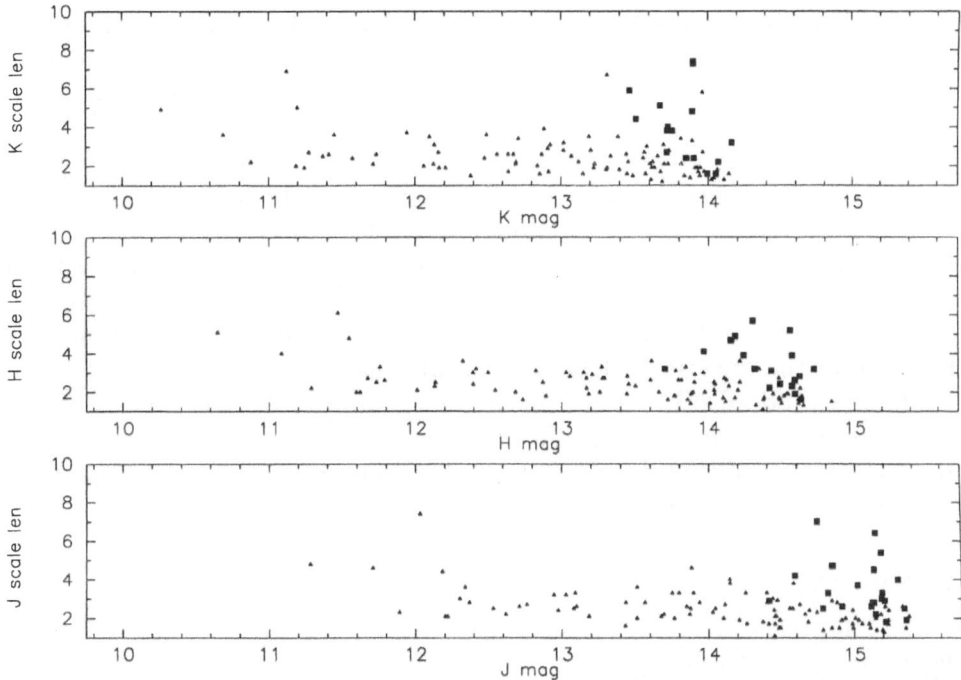

Figure 5. Derived radial profile scale lengths (in arcseconds) for low surface brightness galaxies (large squares) and high surface brightness galaxies (denoted by small triangles).

The results suggest that we are either measuring the "disks" of the LSB galaxies, as opposed to a "spherical" or "bulge" component in which we would expect a profile more similar to a $r^{**}1/4$ law or at least a departure from a simple exponential , or it may suggest a selection effect: 2MASS is more sensitive to spirals than to dwarf ellipticals. In any case, this result is consistent with that seen in the optical for Coma LSB galaxies (Bernstein et al. 1995).

5. Summary

In this study, we examined fields centered on the Coma and SA57 regions (near the north galactic pole). Nearly 20% of all galaxies detected in 2MASS have a central surface brightness faint enough to be classified as type LSB, but most are in the faintest integrated flux bins (near the sensitivity limit of the study) and thus their true number density is incomplete.

Based on the Coma and SA57 preliminary results, the central surface brightness for LSBs appears to be a faint extension of the normal galaxy surface brightness curve. In addition, the J-K color for the LSB galaxies is undistinguishable from that of HSB galaxies. There is no compelling evidence here to suggest that LSB galaxies detected at the 2MASS survey limits belong to a different galaxy population. We caution, however, that the LSB sample is too small to draw any definitive conclusions at this time.

Simple exponential functions are well fit to most radial profiles, suggesting that we are either measuring the "disks" of the LSB galaxies, as opposed to a "spherical" or "bulge" component which we would expect profiles departing from simple exponentials, or it reflects a kind of selection effect – LSBs at the bright limit (of the definition threshold) are well fit by exponentials, regardless of their morphology (i.e., we are not sensitive to the bulge/disk differences). In any case, these results are consistent with that seen in the optical for Coma LSB galaxies.

2MASS (and DENIS) should easily detect low surface brightness galaxies in the local group and the brighter population in clusters at least as far as the Coma cluster ($z = 0.02$). The archive, however, will be far from complete given the sensitivity limits of both all sky surveys.

References

Bernstein, G., Nichol, R., Tyson, J., Ulmer, M., and Wittman, D. 1995, *AJ*, 42, 565.
Bothun, G. *et.al.* 1991, *ApJ*, 376, 404.
Bothun, G. *et.al.* 1993, *AJ*, 106, 530.
Impey, C. *et.al.* 1996, *ApJS*, 105, 209.
Impey, C. and Bothun, G. 1997, *ARAA*, in press.
O'Neil, K., Bothun, G., and Cornell, M. 1997, *AJ*, 113, 1212.
Sandage, A., Binggeli, B. and Tammann, G. 1985, *AJ*, 90, 1759.
Schombert, J. *et.al.* 1992, *AJ*, 103, 1107.
Sprayberry, D. *et.al.* 1996, *ApJ*, 463, 535.
Ulmer, M. *et.al.* 1996, *AJ*, 112, 2517.

2MASS AND THE TULLY-FISHER RELATION

JESSICA L. ROSENBERG

Department of Physics and Astronomy, UMass

1. Introduction

The major advantage of studying galaxies in the near infrared (NIR), is the minimization of dust extinction. The NIR is the region of the stellar emission spectrum which is least affected by both dust emission and extinction. Dust emission becomes more of a factor at longer wavelengths while dust extinction is more significant at shorter ones. The other important advantage of a NIR study is that it probes the bulk of the stellar mass in the galaxies. Since Tully-Fisher (T-F) is a relationship between a galaxy's total stellar content and its rotation speed, minimizing the biases in the measurement of the bulk of the stellar emission provides a significant advantage.

One of the projects that can be carried out with NIR data is determining the peculiar motions of a large number of galaxies. With this information, the derivation of the three dimensional potential and hence the mass density field and the cosmological parameter Ω (Bertschinger *et. al.*, 1990) should be possible. Constraining Ω and studying peculiar motions over the entire sky is one of the core projects for the 2MASS data set.

2. Why Use 2MASS for a Tully-Fisher Study?

One of the 2MASS survey's major advantages over other large T-F studies is sample consistency. 2MASS will provide a uniform sample of galaxies over the entire sky which surpasses the number of sources in all previous catalogs by an order of magnitude. In addition, all of the galaxies observed with 2MASS will also have simultaneous measurements in three NIR bands, J, H, and K. With such a large survey and so many objects to choose from we have the additional advantage that subsamples of the survey will also be large. With the availability of large subsamples, we will be able

N. Epchtein (ed.),
The Impact of Near-Infrared Sky Surveys on Galactic and Extragalactic Astronomy, 247-253.
© 1998 *Kluwer Academic Publishers.*

to apply stringent constraints to the data without harming our sample statistics. There is also a hidden benefit for 2MASS in using the data in a large T-F project; the driving force behind this and the other 2MASS core science projects is database validation. In order to generate a well tested catalog for public consumption, the data must be exercised through careful scientific investigation. A project of this scope requires the use of much of the information generated in the extragalactic database.

One of the most obvious and most basic concerns in any project of this sort is the limitations due to signal-to-noise (S/N). Some of the concerns related to signal-to-noise, such as our ability to measure the disk of the galaxies, will be discussed in later sections. As with any observational project, one has to consider the observational unknowns which include the variability of the point spread function and photometric calibration. These are concerns which will continue to be addressed as the project continues and we obtain more survey data.

Another general concern for any all-sky T-F survey is the effects arising at low Galactic latitude. As we have noted, the NIR minimizes the effects of extinction, but it is still a concern at the lowest Galactic latitudes. In the plane the selection criteria for galaxies is going to be different because of the effect of dust and the fact that we are confusion limited. In particular, more galaxies will be selected at K since it is less affected by extinction than at J which selects most of the objects at high galactic latitude because it is more sensitive.

A more technical concern for the use of the 2MASS survey is how we choose to automate the selection of objects. We have the advantages of a large survey, but with that comes the difficulty of managing the data. We must cull the sample so that we are only analyzing spiral galaxies, since they are the only ones to which the T-F relation applies. We must also eliminate distorted spirals from the data set. Separating morphological types has proved to be a difficult task since it is hard to determine morphology in the NIR. Ideally we would like to have a parameter, or set of parameters, in the database which could be used to distinguish morphological types. To test parameters for this purpose, we have selected a sample of the 2MASS objects and, from the Digitized Sky Survey (DSS) data, have determined the morphology by-eye. It is the by-eye determination which is the only way we have to obtain "truth." It should be noted that even by-eye, using the optical DSS data, it was often difficult to determine a galaxy's morphology. When using this sample, we were not able to find any parameters (we tested many of those in the database, but in this preliminary investigation we did not test them all) which separated out spiral galaxies. If we are not able to find parameters which do distinguish morphological types, we have the option of applying an axis ratio limit to the galaxies used. Since ellipticals do

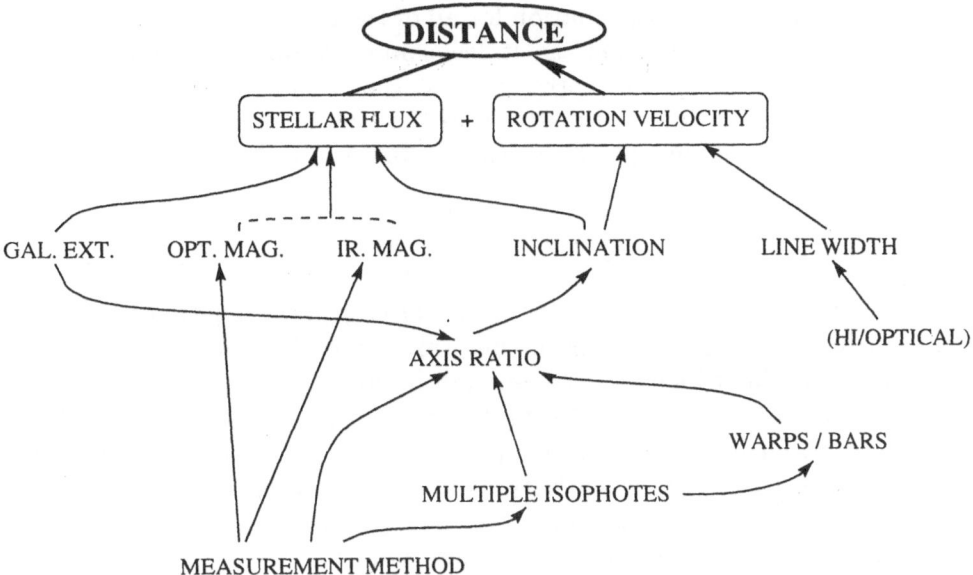

Figure 1. A diagram of the relationship between the parameters which enter into the determination of the T-F distance

not normally have an intrinsic flattening greater than E7, choosing galaxies with axis ratios b/a < 0.3 should guarantee a spiral dominated sample.

3. The Components of the Distance Determination

At the heart of determining mean mass density of the Universe is determining an independent distance to a very large number of galaxies, because individual galaxies may have peculiar motions associated with very localized dynamics, and the accuracy of the T-F method requires large numbers of galaxies in a region to "beat down" the noise. The T-F method to determine a distance requires a stellar flux for a galaxy and its rotation velocity. Figure 1 is a diagram of the interdependence of the various parameters which enter into the derivation of the stellar flux and rotation velocity and hence the distance.

A line width, either optical or HI, is necessary for deriving rotation velocity. The T-F core project will use previously measured line widths and obtain new measurements for galaxies over the entire sky. We have no choice but to obtain additional data in order to determine line widths, but we would prefer to derive the inclination from the existing database. Using optical data would pose the problems of obtaining all of the necessary images, and the impossibility of obtaining those images in the Galactic plane, and potential biases depend on the source of the data.

Deriving an accurate inclination for all of the galaxies is one of the most critical parts of T-F analyses as it feeds into sample selection, rotation velocity determination, and corrections to the stellar flux. We would like to derive the inclination, i, directly from the axis ratio using the standard assumption that the disk has an intrinsic axis ratio (r_o) when seen edge-on so that:

$$cos^2 i = \frac{(\frac{b}{a})^2 - r_o{}^2}{1 - r_o{}^2}$$

where we let $r_o = 0.2$. Peletier & Willner (1991) point out that this is not necessarily the best method for the determination of the inclination, but it is the only one for which we have enough information.

The utility of 2MASS data for finding inclinations obviously depends on our ability to measure the disks of spiral galaxies, which in turn translates into a surface brightness sensitivity. One test of the K-band sensitivity is to compare the axis ratios to optical determinations. In particular we must confirm that we are observing far enough out that we are measuring the disk of the galaxy and that we are not dominated by measurements of the bulge. We have compared the axis ratios as derived at the 3-sigma level in the 2MASS images and the Digitized Sky Survey optical images. If 2MASS is measuring the bulge or is not measuring far enough out in the disk we would expect to be biased in our measurements of the axis ratio relative to the optical values growing worse at small values of b/a. Figure 2 shows the comparison between the NIR and optical measurements of the axis ratios. The lines on the plots are 1-1 lines, not fits to the points. Note that the galaxies represented by the open triangles in the NIR axis ratio (J, H, K, and S which is the super co-add or sum of the three bands) versus visual, V, axis ratio plots are within 0.2 magnitudes of the official survey limit or fainter (most are fainter than the limit). For many of these faint objects, the axis ratio cannot be well determined so a value of 1.0 is assigned. Most of the points follow the 1-1 line with some associated scatter. We will be expanding these comparison samples and testing the accuracy of them for T-F distance comparisons as more data are collected.

The axis ratio of a galaxy is a combination of the galaxy's intrinsic shape and its inclination, but it can also be effected by warps, bars, or the presence of star-formation in the galaxy. An additional advantage of K-band that it is less strongly affected by the star formation regions than are optical wavelengths. Figures 3a and 3b demonstrate the variability of the axis ratio at different points in the K-band images of Messier 64 and 66 respectively. These figures demonstrate how we can be affected by our choice of the position at which we measure the axis ratio. For Messier 64 the vertical lines on the plot represent the 3-sigma and 1-sigma isophotes while for Messier 66 they show the position of the rapid change in axis ratio due

NIR vs. Optical 3—Sigma Axis Ratios

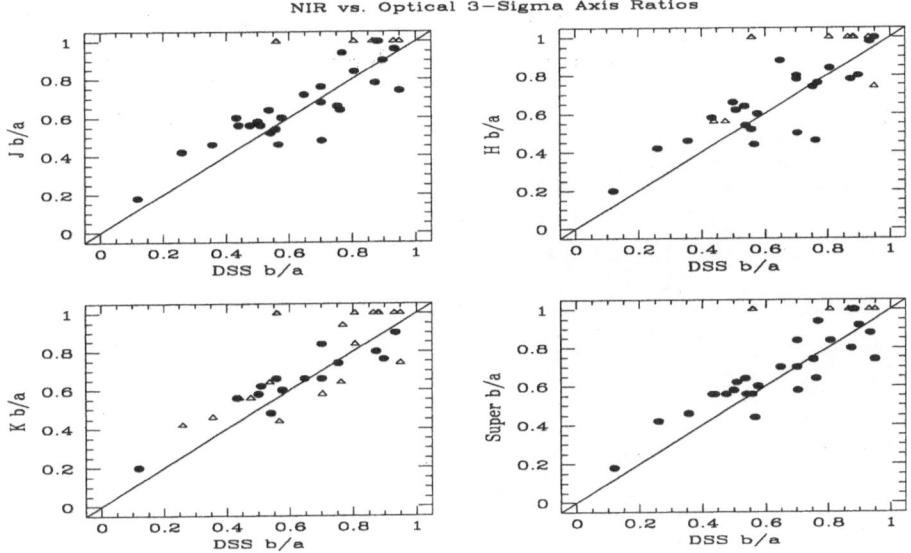

Figure 2. Comparison between NIR and optical measurements of axis ratio determined at the 3-sigma level. The open triangles represent the galaxies whose magnitudes are within 0.2 magnitudes or fainter of the official survey limit. The lines represent the 1-1 relationship between the axis ratios and are not fits to the points.

to the bar, and the 1- and 3-sigma isophotes. The measurement methods need to be studied in detail to determine the most robust axis ratio. We are also testing the use of axis ratio measurements at two separate isophotes for weeding out the warped and irregular objects.

As with determining the axis ratio, we must determine the best way to measure the magnitude for the galaxies. We need to consider whether to use circular magnitudes, which are less biased, or elliptical magnitudes, which are less noisy. In addition, we have to select isophotal magnitudes (and which isophote) extrapolated total magnitudes, or Petrosian magnitudes, or any other type of magnitude. Because of the tight correlation between the T-F parameters, we can invert the relation and minimize the scatter in order to test these magnitudes (Fouqué *et. al.*, 1990).

4. Preliminary Results

We have plunged ahead to make a first stab at testing the 2MASS data using the T-F relation in Hercules. The first caveat is that this was done using prototype camera data. We think that the improvements in the survey camera will improve the data quality and hence these measurements. In addition, all available HI data were used to determine line widths; there

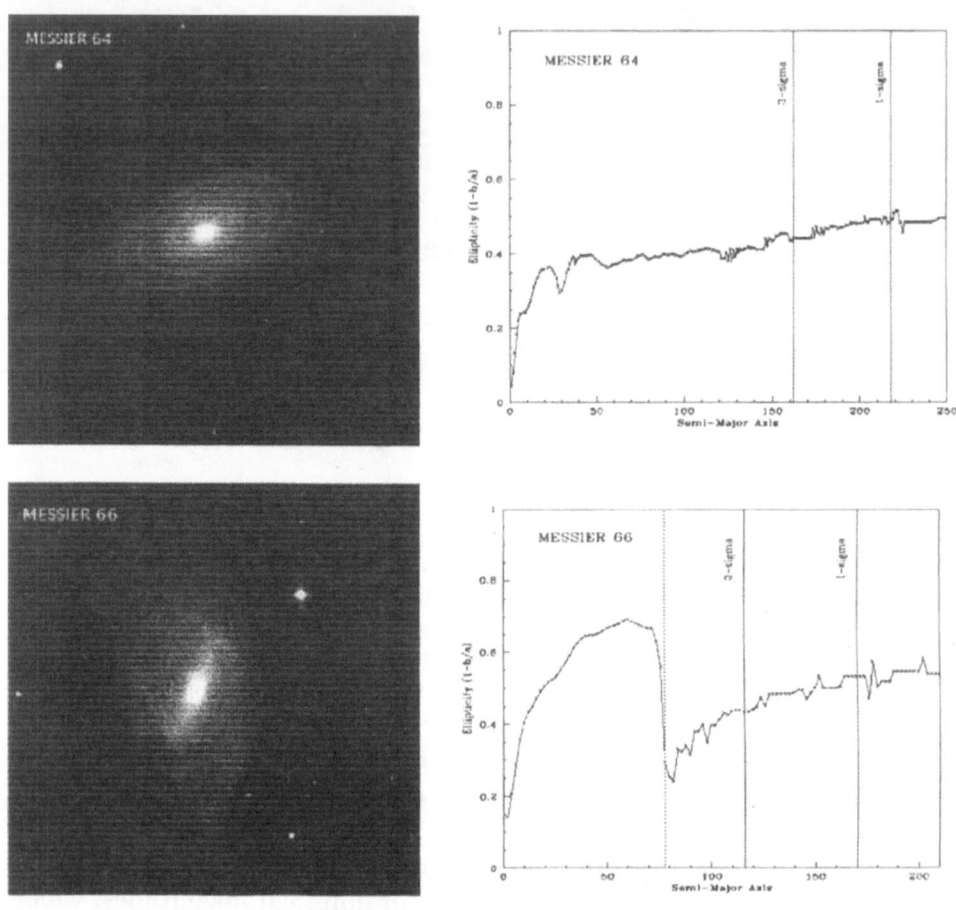

Figure 3.   K-band images and the corresponding plots of ellipticity as a function of position on the semi-major axiw. For M64 we show the axis ratio at the 1- and 3- sigma isophotes. For M66 we show the axis ratio as it changes near the bar as well as the 1- and 3-sigma isophotes

were not enough galaxies to cull the sample and use only the best measurements. To measure the magnitudes of the galaxies we used the K-band elliptical measurements. Some discussion has taken place about whether it is better to use circular or elliptical measurements (Peletier & Willner, 1993; Bothun & Mould, 1987). This magnitude is determined in an aperture which is allowed to grow until the surface brightness in the aperture undergoes and inflection or reaches a predetermined surface brightness cutoff. This is probably not the most robust method for determining the magnitude, but we have not yet tested all of the magnitudes to determine which is best. The relationship between magnitude and the logarithm of the rotation velocity for this data showed a lot of scatter because of the difficulties with the data stated above. Even with the scatter, we do find that the data follow a slope of 9.94 with an RMS of 0.19 when one outlier is eliminated, consistent with previous T-F measurements (eg. Kraan-Korteweg *et. al.*, 1988; Aaronson *et. al.*, 1980; Malhorta *et. al.*, 1996).

5. Summary and Conclusions

There is still much work to be done regarding the testing of the 2MASS parameters for use in the T-F relation. The testing will provide us with the best possible parameters to use for determining distances while also exercising the database. In the end we will improve upon the T-F relation described above and use the result to study peculiar velocities over the entire sky. The result of the mapping will be a better understanding of the large-scale flows in the local Universe, a measurement of the mean mass density, and an improved 2MASS dataset.

References

Aaronson, M., Mould, J., and Huchra, J. 1980, ApJ, **237**, 655

Bertschinger, E., Dekel, A., Faber, S. M., Dressler, A., & Burstein, D. 1990, ApJ, **264**, 370

Bothun, G. D., Mould, J. R. 1987, ApJ, **313**, 629

Fouqué, P., Bottinelli, L., Gouguenheim, L., & Paturel, G., 1990 ApJ, **349**, 1

Kraan-Korteweg, R. C., Cameron, L. M., Tammann, G. A. 1988, ApJ, **331**, 620

Malhorta, S., Spergel, D. N., Rhoads, J. E., Li, Jing 1996, ApJ, **473**, 687

Peletier, R. F. & Willner, S. P. 1991, ApJ, **382**, 382

Peletier, R. F. & Willner, S. P. 1993, ApJ, **418**, 626

QSOS IN THE 2 MICRON ALL-SKY SURVEY

ROC CUTRI, B.O., NELSON, C.J. LONSDALE
Infrared Processing and Analysis Center/JPL Pasadena, USA

AND

H.E. SMITH
Center for Astrophysics and Space Science, UCSD

Abstract. The Two Micron All-Sky Survey (2MASS) will detect QSOs. For example, all of the PG QSOs will be detected by 2MASS in J, H and K_s at a $SNR > 10$. During prototyping observations for 2MASS, 40 QSOs with redshifts in the range $0.1 < z < 3.1$ were serendipitously detected in at least one of the survey bands. Thus, 2MASS will produce a highly uniform set of near infrared photometry for a wide variety radio, optical, ultraviolet and X–ray detected QSOs. 2MASS also has the potential to reveal large numbers of previously unknown AGN. Because QSOs occupy distinct regions of near infrared and infrared-optical color space, it will be possible to carry out highly efficient automated searches of the 2MASS databases in combination with newly available digitized optical sky survey databases for candidates. We review the near IR and optical/IR properties of "conventional" QSOs from UV and optical samples, and estimate the number that will be detected by 2MASS. At minimum, it should be possible to find the southern hemisphere equivalents of the UV-excess PG QSOs (≈ 100 new QSOs in the area $b < 30$ *deg*. There is growing evidence for the existence of a significant population of highly reddened QSOs from IRAS and radio–based surveys (*e.g. Low et al. 1988, ApJ, 327, L41; Webster et al. 1995, Nature, 375, 469; Gregg et al. 1996, AJ, 112, 407*). The infrared and radio surveys suggest that UV–selected searches for QSOs, such as the PG-survey, underestimate by factors of 2-5 the space density of even nearby QSOs. Thus, 2MASS may reveal between several hundred and several thousand new, red QSOs. We discuss 2MASS's ability to test for such new populations of QSOs based on a variety of models of their space density and luminosity function. Finally, we discuss preliminary infrared/infrared-optical color searches and spectroscopic follow-up of new QSO candidates drawn from the 2MASS Prototype Camera database.

N. Epchtein (ed.),
The Impact of Near-Infrared Sky Surveys on Galactic and Extragalactic Astronomy, 255.
© 1998 *Kluwer Academic Publishers.*

THE FIRST-VLA SURVEY:

The First 5000 Square Degrees

R. BECKER
University of California-Davis, and IGPP/LLNL, USA

Abstract.

The VLA FIRST survey now encompasses 5000 sq. deg., mostly in the north Galactic Cap. The First survey generates images of the sky at 1400 MHz with a sensitivity to point sources down to 1 mJy with positions accurate to 1 arcsec (90% confidence). The survey finds approximately 85 discrete sources per sq. degree. Detailed information about the FIRST survey can be obtained at http://sundog.stsci.edu. At the level of the POSS 1, 15% of FIRST sources have optical counterparts. While these are primarily galaxies, there are significant numbers of quasars and stars detected in the survey. This presentation will highlight some of the research being pursued by the FIRST science team.

N. Epchtein (ed.),
The Impact of Near-Infrared Sky Surveys on Galactic and Extragalactic Astronomy, 257.
© 1998 *Kluwer Academic Publishers.*

FIRST-VLA RADIOSOURCES IN ELAIS AREAS

E. A. GONZÁLEZ–SOLARES AND I. PÉREZ-FOURNON
Instituto de Astrofísica de Canarias
C/ Via Lactea
38200 La Laguna
Tenerife, Spain

AND

ELAIS COLLABORATION

Abstract. ELAIS (*European Large Area ISO Survey*) is a project that has carried out a deep wide angle survey with ISO at 7, 15 and 90 μm over an area of $\sim$13 square degrees of high latitude sky. The value of the ELAIS survey will be enhanced by carrying photometry at 7 wavelengths on $\sim$300 sources at 4.5, 6.7, 7.8, 15, 25, 50, 90, 135 and 180 microns. We present a cross-correlation of the ELAIS sources and the FIRST-VLA radio sources in the CAM test survey area.

1. Introduction

The Infrared Astronomical Satellite (IRAS) had enormous success arising principally from its survey products (particularly the Point Source Catalog and the Faint Source Catalog). Perhaps most significant was the discovery of a whole new class of objects with enormously high far infrared luminosity [notably F10214+4724 (Rowan–Robinson *et al.*, 1991a) and P09104+4109 (Kleinmann *et al.*, 1988)]. As well as discovering new objects, IRAS demonstrated the benefit of selecting objects in the far infrared. This wave band is not sensitive to dust obscuration which biases optically selected samples. The emission arises from thermally heated dust and thus complements studies of emission directly from starlight, gas or AGN engines.

The Infrared Space Observatory (ISO) is the successor to IRAS and is providing unparalled observations in the mid and far infrared. The sensitivity of ISO is orders of magnitude better than IRAS.

259

N. Epchtein (ed.),
The Impact of Near-Infrared Sky Surveys on Galactic and Extragalactic Astronomy, 259-263.
© 1998 *Kluwer Academic Publishers.*

2. ELAIS survey

The European Large Area ISO Survey (ELAIS) is a project that has now surveyed ~13 square degrees of the sky at 15 and 90 μm and ~7 square degrees at 6.7μm with the Infrared Space Observatory (ISO). ISO photometry will be carried out on around 300 sources.

The ELAIS survey areas have been carefully selected to be at high galactic latitude and also in regions of the lowest dust column density and cirrus emission. The survey regions consist of 4 regions ~ 2° × 2° and 6 smaller regions 20′ × 20′ in extent (see table 1).

The choice of where to distribute these areas on the sky was governed by a number of factors. Firstly we decided not to group these all in a single contiguous region of the sky. Had we done so, we may have difficulty distinguishing evolutionary effects from local large scale structures. Cirrus confusion is a particular problem, so we selected regions with low IRAS 100 μm intensities using the maps of (Rowan–Robinson *et al.*, 1991b). To avoid conflict with other ISO observations we further restricted ourselves to regions of high visibility over the mission lifetime (> 25%). To avoid high zodiacal backgrounds we only selected regions with high ecliptic latitudes ($|\beta| > 40°$). Finally it was essential to avoid saturation of the CAM detectors so we had to avoid any bright IRAS source.

At our survey limit ~ 50 mJy at 90 μm, ISO may be confusion limited by galaxies and galactic cirrus emission and hence our survey will be the deepest far infrared survey possible with the ISO satellite. The 15 μm survey shall reach a 5 sigma sensitivity of 2 mJy. The survey is detecting objects 5-10 times fainter than IRAS in the 50-100 μm range and 20-50 times fainter than IRAS in the 10-20 μm range.

2.1. ELAIS SCIENTIFIC AIMS

While it is impossible to predict all the scientific benefits of such a large project, below is outlined some of the key issues that we hope to address. A major theme is the detection of high redshift galaxies, the derivation of the star formation history of the Universe between now and a redshift 1 and the compilation of unbiased samples of active galaxies.

2.1.1. *Epoch of Galaxy Formation*

Ly-alpha searches have failed to find the epoch of galaxy formation. If elliptical galaxies underwent an intense period of star formation accompanied by prodigious quantities of dust they may be undetectable in the optical but detectable in the far infrared

2.1.2. *Star Formation in Spiral Galaxies at High Redshift*
The main extra-galactic population detected by IRAS was galaxies with high rates of star formation. The far infrared emission arises from dust heated by young stellar populations. The sensitivity of ISO will allow us to detect these objects at much higher redshifts and thus abtain greater understanding of the cosmological evolution of star formation.

2.1.3. *Ultra and Hyper-Luminous IR Galaxies at high redshift*
IRAS uncovered a population with enormous far infrared luminosities. Exploration of these objects at higher redshift will have particular significance for models of AGN/galaxy evolution.

2.1.4. *Emission from dusty tori around AGN*
Unified models of AGN suggest that the central engine is surrounded by a dusty torus. Optical properties are then dependent on the inclination angle of this torus. The far infrared emission from the torus will be less sensitive to the viewing angle. The far infrared properties of these will place important constraints on unification schemes.

2.1.5. *Dust in normal Galaxies to Cosmological Distances*
Faint optical redshift surveys find surprisingly few galaxies beyond $z=1$. One possible explanation for this is a dust fraction that increases with z. Emission from the cool interstellar 'cirrus' dust in normal galaxies will be detectable in our survey to much grater distances than were accesible with IRAS, so we will be able to examine the dust content to higher z.

2.1.6. *Circumstellar Dust Emission from Galactic Halo Stars*
The deep stellar number counts provided by this survey will be relatively unaffected by Galactic extinction and may provide improved estimates of the halo/disk population ratios.

2.1.7. *New classes of objects*
F10214 was discovered at the limit of the IRAS capability and it is reasonable to hope that equally unexpected and exciting objects will be discovered at the limits of this survey.

2.1.8. *Clustering Properties*
The volume of this survey is comparable to that surveyed by the entire IRAS Point Source Catalog. The median redshift will be much higher. We will thus be in a position to examine the evolution of clustering strength, giving perhaps the most direct test of gravitational instability picture of structure formation.

TABLE 1. Summary of Areas. The first four areas comprise the main survey made up from $40' \times 40'$ rasters. One raster in N3 will be repeated. The final 6 areas are single smaller rasters $20' \times 20'$

Area	Rasters	Nominal Coordinates J2000		$\langle I_{100} \rangle$ /MJysr^{-1}	Visibility /%	β
N1	3×3	$16^h 08^m 44^s$	$+56°26'30''$	1.2	98.0	73
N2	4×2	$16^h 39^m 34^s$	$+41°15'34''$	1.1	58.7	62
N3	3×3	$14^h 28^m 26^s$	$+32°25'13''$	0.9	26.9	45
S1	4×3	$00^h 38^m 24^s$	$-43°32'02''$	1.1	32.4	-43
Lock. 3	1	$13^h 34^m 36^s$	$+37°54'36''$	0.9	17.3	44
Sculptor	1	$00^h 22^m 48^s$	$-30°06'30''$	1.3	27.5	-30
TX1436	1	$14^h 36^m 43^s$	$+15°44'13''$	1.7	22.2	29
4C24.28	1	$13^h 48^m 15^s$	$+24°15'50''$	1.4	16.8	33
VLA 8	1	$17^h 14^m 14^s$	$+50°15'24''$	2.0	99.8	73
Phoenix	1	$01^h 13^m 13^s$	$-45°14'07''$	1.4	36	

3. Needs for radio data

The spatial resolution of ISO is insufficient to unambiguously identify many of the faintest ISO sources. Even at 15 μm, the survey resolution is $\sim 10''$ and at 90 μm it is $\sim 2'$ so there will be multiple optical counterparts within each error ellipse. Complementary radio data will play a crucial role in identifying many of the most interesting objects.

In addition, combining deep radio and optical data with the ISO survey fluxes will provide information on the trivariate IR-radio-optical luminosity function and its evolution and the contribution of starburst galaxies to the sub-mJy radio source counts. The ratio of the FIR emission and radio emission will also allow us to investigate the physical origin and spatial distribution of the energy sources in the detected objects in the same way that the VLA maps have been central to our understanding of the origin of IRAS sources.

We expect that most of the ISO sources will be detected at the 0.1-0.5 mJy level at 20 cm. The physical basis for this is the well established FIR/radio correlation (Helou et al., 1985).

3.1. THE FIRST SURVEY

The VLA FIRST (*Faint Image of the Radio Sky at Twenty-cm*) survey is a survey of the sky obtained with the VLA at a wavelength of 20 cm. Images are produced with 1.8'' a typical rms of 0.15 mJy and a resolution of 5''

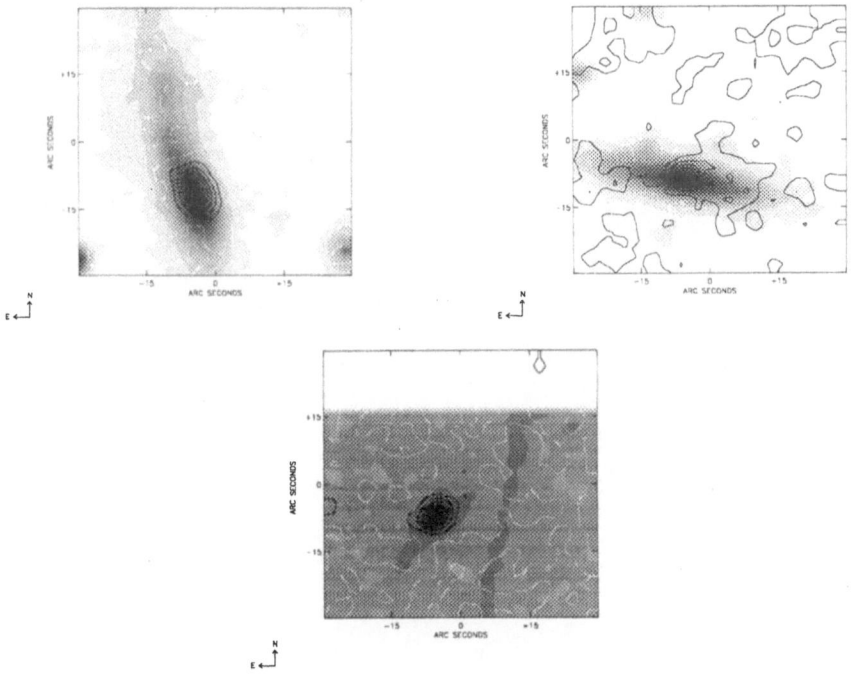

Figure 1. FIRST radio sources with 15μm counterpart in the ELAIS CAM test region. Radio contours are plotted over an optical DSS image.

(Becker *et al.*, 1994).

The ELAIS test region is a 43′ × 43′ area in N2 centered on $\alpha(2000) = $ 16 35 45, $\delta(2000) = $ +41 06 00. This region was the first one observed with CAM and the first one to be analyzed.

The ISO data at 15 μm of the CAM test region has 53 sources detected with a nominal reliability of 95% and a flux limit 0.5 mJy. There are 94 FIRST radio sources in ELAIS CAM test raster, 20 of them are point-like and 74 extended. Only 3 of them are also detected at $15\mu m$ (2 are known IRAS galaxies). Figure 1 shows optical images of these 3 sources with radio contours from FIRST.

References

Becker, R.H. *et al.*, 1994, ApJ, **450**, 559
Helou *et al.*, 1985, ApJ, **298**, L7-L11
Kleinmann *el al.*, 1988, ApJ, **328**, 161
Rowan–Robinson *et al.*, 1991a, Nature, **351**, 719–721
Rowan–Robinson *et al.*, 1991b, Mon. Not. R. Astron. Soc., **249**, 729–741

VI- Miscellaneous

7.4 Miscellaneous

THE SOM-FITTING METHOD

E. BERTIN
Sterrewacht Leiden
(presently at ESO-Garching, Germany)

Abstract. We present a new astronomical image analysis method based on Self-Organizing Maps (SOMs). The SOM-fitting method allows one to perform accurate photometry, astrometry and classification using a morphologically-ordered set of codebook image patterns. The prototype software and some preliminary results on DeNIS simulations and real survey images are presented.

1. Introduction

Profile-fitting has proven to be one of the most accurate methods for deriving astrometrical and photometrical measurements of sources in astronomical images. When the Point Spread Function (PSF) can be considered as constant, it is fairly easy to apply profile-fitting to stellar fields. Comparison with the PSF can also provide an efficient star/galaxy classification.

Unfortunately, in large, ground-based imaging surveys one often has to deal with a PSF that is strongly variable with time and position. Variations with time may originate from changes in the seeing and/or uncompensated mechanical flexures. Variations with position, due to imperfect optics, are very common on large-field instruments. Among other aberrations, coma is then generally most obvious to the eye. Figure 1 shows two samples of the DeNIS instrumental PSF from the same K-band image. How shall we handle such sources?

In all DeNIS bands, aperture-magnitude growth curves prove to be stable (at the 1% level) for radii $\gtrsim 5$". Such aperture magnitudes can therefore be used to normalize the PSF, but will not be optimum for measuring directly faint stars (contribution of the background noise) or in crowded fields (contamination by neighbours). On the other hand, fitting with an inadequate profile yields significant photometric biases: up to $\approx 10-20\%$ for the DeNIS PSF variations. These biases are not easy to handle, as they may not

N. Epchtein (ed.),
The Impact of Near-Infrared Sky Surveys on Galactic and Extragalactic Astronomy, 267-276.
© 1998 *Kluwer Academic Publishers.*

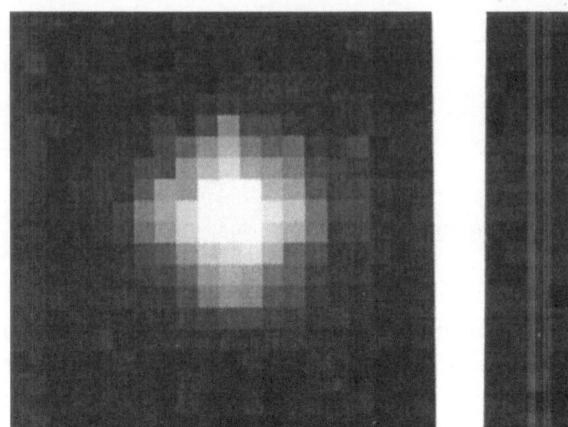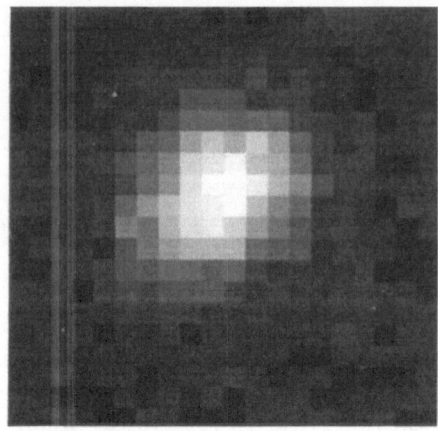

Figure 1. Left: a star near the center of a DeNIS K frame. Right: in a frame's corner.

only depend on the position of the source in the image, but also on its flux (the pixel-to-pixel ratios of appropriate weights are a function of the flux in the régime of the DeNIS image noise model). In addition, star/galaxy separation needs proper modeling of the PSF (see Mamon, these proceedings), which encourages us to search for a global method for representing the shapes of sources as they appear throughout the images.

2. General concept

Looking at any astronomical image, it is obvious that detections — to a certain degree of precision — come in a fairly limited number of shapes. Let's suppose that we find suitable prototypes for representing most of these objects. We may sort these prototypes in some low-dimensional space in such a way that we can locally interpolate them to give a "fair" representation of most sources that we may encounter in our image (or set of images). We might then assign to any detected and catalogued source a "feature vector" that gives the position of the source in this "feature-space". Doing so, we have achieved some kind of lossy image compression by providing a mapping from pixel-space to feature-space. We should then be able to use this low-dimensional feature-map as simply as, for instance, a colour-magnitude diagram, with the possibility to isolate groups of objects that share similar morphologies, study correlation with position, etc.

3. Self-Organizing Maps (SOM)

The Self-Organizing Map algorithm (Kohonen 1987) is precisely intended to provide such a mapping from some input space to a feature space. SOMs, as a neural network model, share similarities with the organization of neural functions found in the brain. I will only describe here the basics of the method and my own modifications done to it. See Kohonen (1997) for more details. Some preliminary attempts of classification of astronomical images with SOMs can be found in Mähönen & Hakala (1995) or Naim et al. (1997).

3.1. THE BASIC ALGORITHM

Each prototype i is a node (sometimes called neuron) of a lattice which can be given different topologies; for simplicity we have adopted here (hyper)cubic meshes. Each node is assigned a characteristic vector $\mathbf{m}_i$, initialized to some random direction and module. "Learning" will consist in adjusting iteratively and both competitively and cooperatively these weight vectors so that they converge to a state where they provide a (self-) organized map of prototype vectors for a given set of input data.

Basically, each iteration consists of picking a new input vector $\mathbf{x}$ and computing for each node the Euclidean distance

$$d_i = ||\mathbf{x} - \mathbf{m}_i|| \tag{1}$$

A "winning" node c with the lowest distance d_c is then found and node vectors are updated according to

$$\mathbf{m}_i(t+1) = \mathbf{m}_i(t) + h_{ci}(t).(\mathbf{x} - \mathbf{m}_c), \tag{2}$$

where h_{ci} is a kernel function which specifies the domain around the winning node where the update should be made. A common choice is the gaussian kernel:

$$h_{ci} = \eta(t)e^{-(r_c-r_i)^2/2\sigma^2(t)}, \tag{3}$$

where $\eta(t)$ and $\sigma(t)$ specify, respectively, the learning rate and the kernel width (both decreasing functions of time). Because of the finite range of h_{ci}, "pockets" rapidly emerge in the lattice, where groups of similar patterns can be found. These groups tend to organize their relative positions in such a manner that a large-scale ordering appears. As $\eta(t)$ and $\sigma(t)$ decrease, the system slowly "freezes" in a state where nodes have become faithful prototypes of input patterns. The SOM algorithm indeed tends to minimize the so-called "average distortion measure":

$$\sum_j \sum_i h_{c(j)i}||\mathbf{x_j} - \mathbf{m}_i|| \tag{4}$$

3.2. MODIFICATIONS

For astronomical purposes, several important modifications had to be made
to the original SOM algorithm. In particular, handling of *weighted* input
vector components was added; and instead of normalizing the *variance* of
input vectors (which is done, when possible, to speed-up convergence), we
normalize their modulus.

3.3. DIMENSIONS OF THE SOM

Before learning, the number D of dimensions, and the number of nodes
along each dimension have to be specified. For the examples shown here
$D = 2$ is used, with 5 to 20 nodes per axis. For large practical applications
featuring strongly variable PSFs and galaxies, it is better to use $D = 3$.
This avoids a "folding" of the map in the input space.

4. Input and output spaces

What information should be put in the input pattern vectors to describe the
detections? Although there are many interesting ways of describing a profile
(fitted sets of parameters, principal components,...), the current choice is to
use directly an array of pixel values centered around each detection. This
has many advantages: linearity (no bias when S/N becomes low); weights
are easy to assign to each component (bad or saturated pixels for instance
get zero weight); any shape can be handled; all the information up to a
given scale is kept, etc... The main problems are that: (1) only partial
information is available for very extended objects, and (2) the patterns
are not shift-invariant, even for well-sampled images. Problem (1) is not
critical for the DeNIS sources because the main interest lies in stars and
faint galaxies. Problem (2) is bypassed by recentering each detection prior to
learning, using sinc-interpolation. Subtraction of a Gaussian core is applied
to remove most of the aliased component when dealing with moderately
undersampled data (1 pixel < FWHM < 2.3 pixels). After learning, two
extra-dimensions are added to the "pure" feature-map, with prototypes
regularly resampled over the pixel grid using an appropriate step[1]. Figure
2 shows the content of a typical SOM.

---

[1] Any Gaussian feature can be recovered through linear interpolation, with a maximum
fractional error $< \epsilon$, if the step used is smaller than FWHM $\times \sqrt{\epsilon / \ln 2}$.

Figure 2.   Small 2-dimensional feature-map with 7 × 7 nodes, produced after training on a CCD image. Each prototype vector is represented here in grey levels (logarithmic scale). Note the "cosmic-ray" (lower right) and "bad column" (lower left) prototypes. Galaxy shapes can be found in the upper left part of the array.

5. SOM-fitting

5.1. IMPLEMENTATION

The prototype of a "SOM-fitting" code has been implemented in the SExtractor detection software (Bertin & Arnouts 1996). Once learning is accomplished (on a good workstation this generally takes no more than a few minutes for several thousands of sources), the module containing the m_i is loaded in SExtractor and used to interpret each source detected on the current images. Given a source, one can define its "error landscape" in feature-space: a χ^2 estimator based on expected photon-statistics defines the intensity of the error as a function of position in feature (+position) space. A typical error landscape, in two feature dimensions, is shown in Fig. 3. As it can be seen, a minimum with a reduced error of ≈ 1 can be found for this object. Currently, the χ^2 minimization is done through a combination of grid-search and conjugate-gradient methods over the linearly-interpolated grid of prototypes.

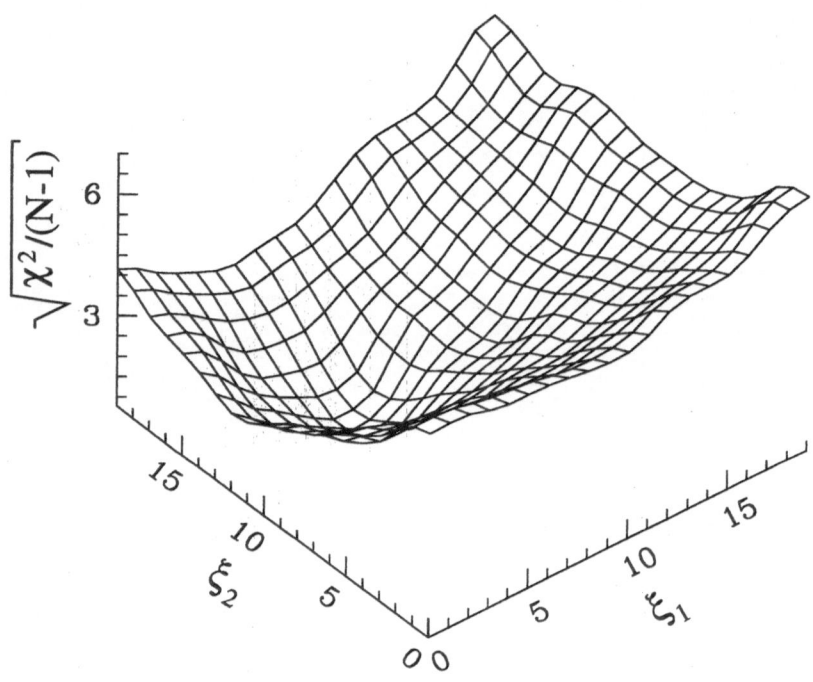

Figure 3. Typical χ^2 error landscape in a simple case. Flux is a free parameter, which explains the "$\chi^2/(N-1)$" label for the vertical axis. The ξ_1 and ξ_2 axes are feature-space coordinates. The two resampling coordinates have been dropped here for clarity.

5.2. CROWDED FIELDS

In I-band DeNIS images where the source density is low, SOM-fitting generally gives reduced errors close to 1, even for fairly bright stars (in the microscanned J and K channels, this is not so good because of the jitter of the atmospheric PSF between the short sub-exposures). In crowded fields the situation degrades significantly, because of blending. One may therefore occasionally allow for n prototypes to be fitted simultaneously, which implies adjusting n times more parameters. As a first step for the prototype considered here, the program limits itself to two passes through the data, one to remove the best-fitting prototypes; and a second one where each prototype is temporarily added back and the fitting done again. As Fig. 4 shows, this method already yields fairly good results in crowded fields. In the meanwhile, SOM-fitting provides position estimates of prototypes which are much more accurate than usual barycentering: improvements in astrometric precision of about 4× are not uncommon.

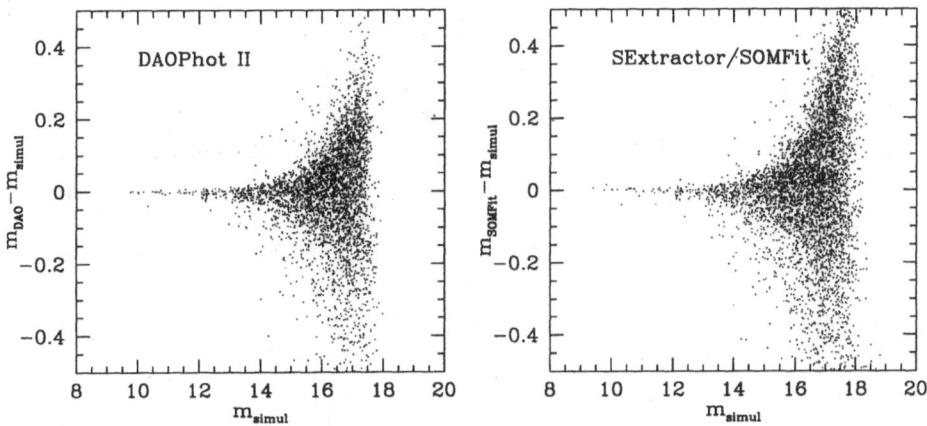

Figure 4.   Photometric comparison between DAOphot2's "allframe" and SExtractor's SOMfit prototype on a DeNIS-I band simulated image of a galactic bulge field produced with SkyMaker2[2]. The seeing FWHM is 2 pixels. With the detection parameters set here for DAOphot and SExtractor, the latter detects 30% more stars, mostly on the faint side. Note the trend seen on both plots towards faint magnitudes: it can be explained as an artifact of the cross-identification process with the simulated catalog, due to the strong density of background sources.

5.3. VARIABLE PSFS

From what has been seen so far, the handling of variable PSFs with SOMs should be straightforward. Indeed, as Fig. 5 demonstrates, the SOM-fitting method maps PSF distorsions along the field in a rather efficient way. What is needed now is a mapping from x, y positions (and possibly time t) to

---

[2]ftp://ftp.iap.fr/pub/from_users/bertin/skymaker/

feature-space. In the current prototype, this is done by adding *explicitely* x, y to the input vector. This has, however, the disadvantage of generating unnecessary prototypes, especially for image zones where the PSF stays stable. Tests are currently underway to find a more general method which would also be able to deal with brutal PSF changes, like those that can be found in coadded images.

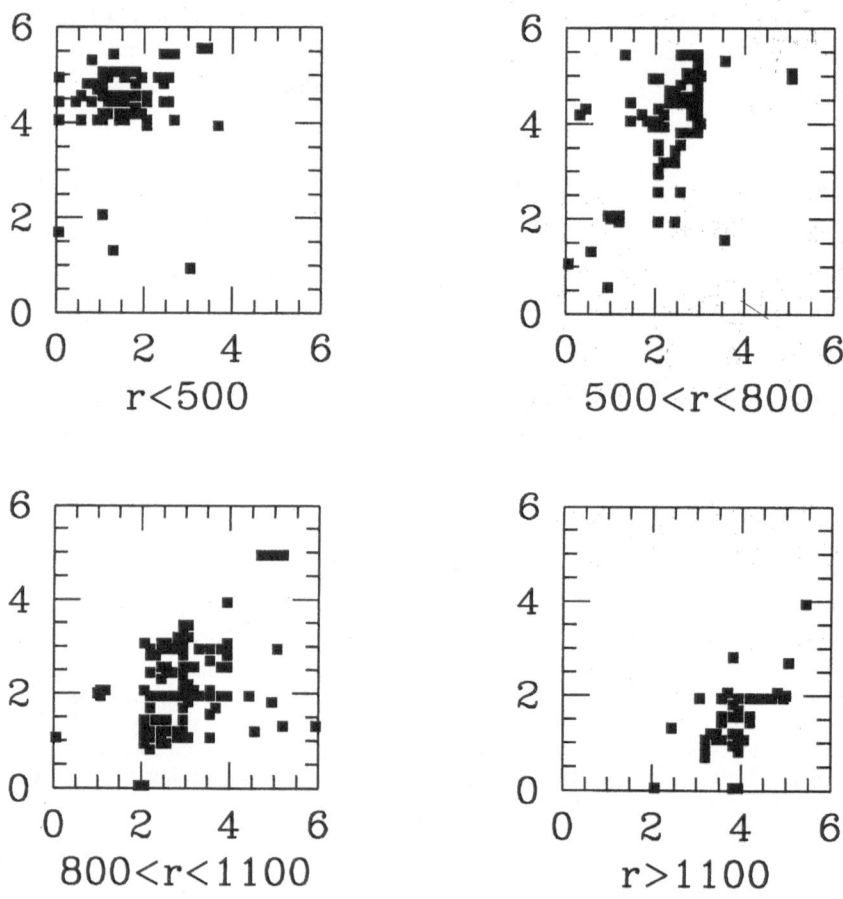

Figure 5. (Two-dimensional) feature-space coordinates of sources extracted in 4 different zones of an image with variable PSF. The image is a simulated DeNIS I-band, crowded, stellar field to which a variable coma aberration component has been added. The purely "atmospheric" FWHM is 1.5 pixels, and the d80 (diameter of the disk within which 80% of the light is enclosed) of the coma component varies radially across the field from 0 (at radius $r = 0$) to 2 pixels ($r = 1500$). Only stars with I< 13 are plotted.

5.4. CLASSIFICATION AND STAR/GALAXY SEPARATION

Figures 6 and 7 show how different kind of objects can be selected in a feature map. It is important to stress that the SOMfitting method is not more than an abstraction tool for classification: all it does is reducing dramatically the dimensionality of pattern vectors. A discriminant analysis is still required to perform the classification itself. A star/galaxy separation method based on SOM-fitting is currently under development. Preliminary tests indicate a net increase in performance in comparison to SExtractor's CLASS_STAR classifier, especially in crowded regions, which is of prime importance for studies of the Zone of Avoidance.

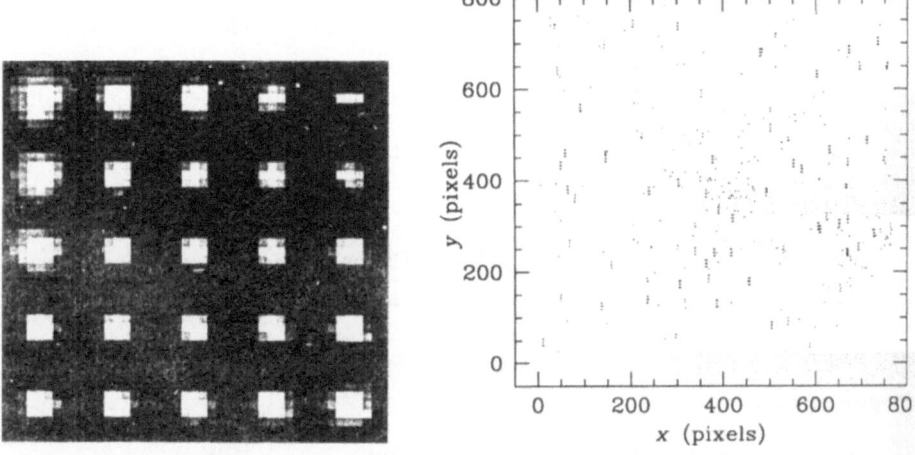

Figure 6. Left: Typical feature map obtained for a DeNIS J-band image. Right: x, y positions, for a full DeNIS strip, of sources that were identified as the upper rightmost prototype in the same feature map. Note the peculiar arrangement of bad pixels, due to microscanning.

6. Conclusion and perspectives

We have shown that SOM-fitting is a promising tool for handling morphological information in imaging surveys. Photometry and astrometry, as well as classification are likely to benefit from this new technique. Work is currently underway to implement it as a module ("SOMfit") in the SExtractor package. Many improvements are still necessary to make the system fully general, though. This includes the handling of large objects. In addition, only shift- and flux-invariances are implemented yet. Rotation and possibly scale invariances would be of great interest for analysing data with a high image quality (isotropic PSF, highly-resolved sources).

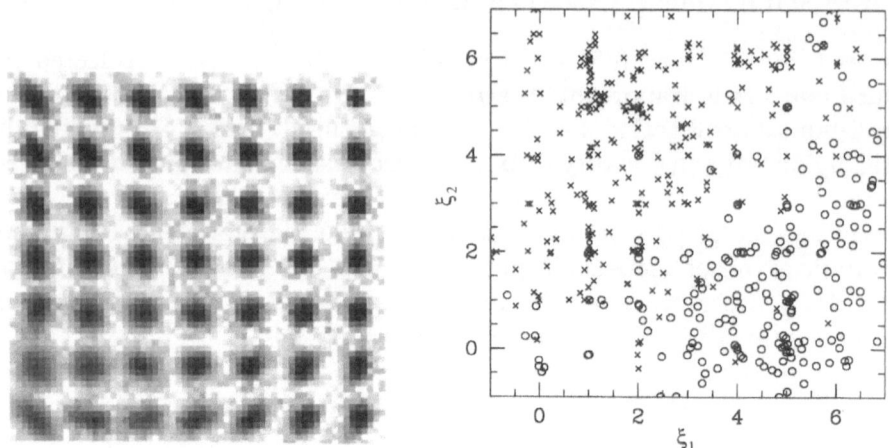

Figure 7. Left: Feature map obtained from a deep NTT image (negative, logarithmic scale). Note the prefered orientation of elongated galaxy prototypes along the diagonals, due to the square shape of the selection-window. Right: position, in the same feature-space, of sources classified with SExtractor's CLASS_STAR parameter as stars (crosses) and galaxies (circles).

Acknowledgments   I would like to thank Pascal Fouqué for interesting discussions and Erik Deul for helpful suggestions made on the manuscript of this paper.

References

Bertin E., Arnouts S. 1996, *A&AS*, **117**, 393

Kohonen T., 1987, *Self-Organization and Associative Memory*, 2nd edition, Springer

Kohonen T., 1997, *Self-Organizing Maps*, 2nd edition, Springer

Mähönen P.H., Hakala P.J., 1995, *ApJ* **L77**, 452

Naim A., Ratnatunga K.U., Griffiths R.E., 1997, *ApJS* **111**, 357

NEAR INFRARED OBSERVATIONS OF ASTEROIDS

TO IMPROVE THE ASTEROID CLASSIFICATIONS.

A. BAUDRAND
Bureau des longitudes, URA CNRS 707, Paris
Observatoire de Paris, Meudon, DESPA

A. BEC-BORSENBERGER
Bureau des longitudes, URA CNRS 707, Paris

A. BARUCCI
Observatoire de Paris, Meudon, DESPA

AND

G. SIMON
Observatoire de Paris, DASGAL

Abstract. In the frame of the Deep Near Infrared southern sky Survey (DENIS), our program is to recover all the asteroids observed during the survey and to determine their I, J and K magnitudes. The aim of our work is to enlarge the asteroid colorimetry data base to the near-infrared in order to improve the knowledge of the compositional structure of the asteroid population.

1. Introduction

Classification allows to investigate the compositional structure of the whole asteroid population. A better knowledge of the chemical composition of asteroids can help in understanding the origin and the evolution of this population. These small bodies contain a large amount of information regarding some of the primordial processes which governed the evolution of the Solar System, immediately after the collapse of the protoplanetary nebula and before the formation of planets. Small differences between objects imply different histories (undisturbed survival, differenciation, collisional fragmentation...). Therefore, each taxonomic unit may record the final stages of one of these histories. It follows that a good model for the evolution of the Solar

N. Epchtein (ed.),
The Impact of Near-Infrared Sky Surveys on Galactic and Extragalactic Astronomy, 277-280.
© 1998 *Kluwer Academic Publishers.*

System has to take into account, as boundary conditions, the multiplicity of asteroid types and their physical implications.

The recent classifications of asteroids obtained by means of statistical analysis (Tholen 1984, Barucci et al. 1987 and Tedesco et al. 1989) are based essentially on colorimetry over the wavelength range of 0.3 to 1.1 microns and the albedo obtained by IRAS (Tedesco et al. 1992). The asteroids have been grouped according to their spectral characteristics in several classes which are connected to heliocentric distance and compared to meteorites.

Enlarging the asteroid colorimetry data base to the near infrared field will allow to improve the existing classification. Indeed, the infrared colors add new dimensions to the parameter space used, which may lead to refine the distinctions between existing classes or even to recognize new ones.

2. Utilization of DENIS data and first results

At present, 2000 strips (a strip is composed of 180 images of 12'x12') have been observed. The possibility of retrieving asteroids in the strips has been studied. It can be envisaged in two different ways using the DENIS data base called FOURBI, which provides information on the observed areas characterized by their geocentric equatorial coordinates and their epoch of observation:

1) when the position of one given asteroid at a given epoch is known, we search the strips in which observed areas correspond to the calculated one;

2) more generally, we have to determine the asteroids which are in each observed area. For this, we have to create a good interface between asteroid ephemerides and the FOURBI data base.

In this preliminary study, we have considered only the first point and we have obtained the magnitudes of the following asteroids: 253 Mathilde, 423 Diotima, 669 Kypria, 633 Zelima, and 1515 Perrotin, as reported in the Table 1.

TABLE 1. Apparent magnitudes (* beyond DENIS limits)

Asteroid	Epoch	I	J	K
253	96-03-26	14.29 ± 0.02	14.11 ± 0.04	13.54 ± 0.29
423	96-03-27	not observed	11.10 ± 0.03	10.57 ± 0.03
633	96-03-28	14.03 ± 0.03	13.58 ± 0.04	13.13 ± 0.15
669	96-03-26	14.21 ± 0.07	15.01 ± 0.48	*
1515	96-03-29	16.94	*	*

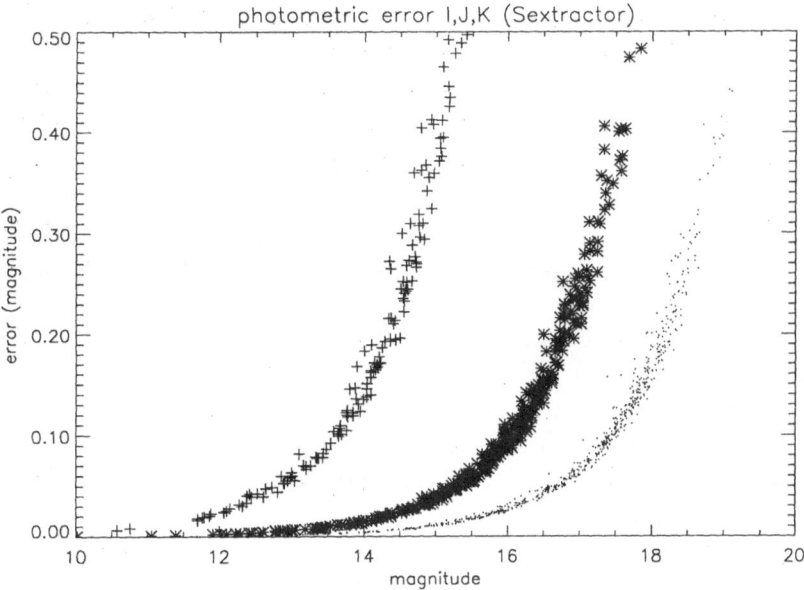

Figure 1. Limits of the three filters with the relative errors. Points, stars, crosses mark the filters I, J and K, respectively.

As it is showed by the Table 1, the apparent magnitude of some observed asteroids will be beyond the DENIS limits. Therefore we will have to select the most favorable period of observations. Figure 1 shows the limits of the three filters with the relative errors.

3. Conclusion

The statistical method of classification that will be used, called the G-mode method (Coradini et al., 1977), allows the user to obtain an automatic classification of the asteroids in spectrally homogeneous groups. The role of infrared colours in separating the various groups has already been studied (Birlan et al., 1996) and the results obtained with IR data are encouraging for continuing the observations in this region.

Up to now, the asteroid investigation in the near infrared is still poor. Therefore, the DENIS survey is a good opportunity to get such information, especially as it offers a large and homogeneous amount of data. Indeed, homogeneity and largeness of the data base used are both crucial qualities for all the statistical classifications.

In this way, the DENIS contribution to the asteroid classifications will be both quantitative and qualitative. On one hand, it will increase the number of asteroids used. On the other hand, the increase of the number of measured parameters describing the sample will better describe the nature of the studied objects.

References

Barucci M. A., Capria M. T., Coradini A. and Fulchignoni M., 1987, Icarus, 72, 304-324.

Birlan M., Barucci M. A. and Fulchignoni M., 1996, A&A, 305,984-988.

Coradini A., Fulchignoni M., Fanucci O. and Gavrishin A. I., 1977, Comput. Geosci. 3, 85.

Tedesco E., Williams J. G., Matson D. L., Veeder G. J., Gradie J. C. et al., 1989, Astronom. J., 97, 580.

Tedesco E., Veeder G. J., Fowler J. W., Chillemi J. R., 1992, IRAS Minor planets survey-final report, Phillips Laboratory.

Tholen D. J., 1984, Ph. D. Asteroid taxonomy from cluster analysis of photometry, Doctoral thesis, Univ. of Arizona.

INDEX of AUTHORS